Forest Development

Springer-Verlag Berlin Heidelberg GmbH

Achim Dohrenbusch · Norbert Bartsch
(Eds.)

Forest Development

Succession, Environmental Stress and Forest Management

Case Studies

With 90 Figures and 50 Tables

Editors:

Professor Dr. Achim Dohrenbusch
University of Göttingen
Institute of Silviculture
Büsgenweg 1
37077 Göttingen
Germany
E-mail: *adohren@gwdgt.de*

Dr . Norbert Bartsch
University of Göttingen
Institute of Silviculture
Büsgenweg 1
37077 Göttingen
Germany
E-mail: *n.bartsch@forst.uni-goettingen.de*

ISBN 978-3-642-62857-3   ISBN 978-3-642-55663-0 (eBook)
DOI 10.1007/978-3-642-55663-0

Library of Congress Cataloging-in-Publication Data
Forest development: succession, environmental stress, and forest management / Achim Dohrenbusch, Norbert Bartsch (eds.). p. cm. Includes bibliographical references. ISBN 978-3-642-62857-3   1.

Forest management--Germany. 2. Forest ecology--Germany. 3. Forest management. 4. Forest ecology. I. Dohrenbusch, A. (Achim), 1955- II. Bartsch, Norbert, 1955-
SD195 .F65 2002   333.75'0943--dc21   2002030231

http://www.springer.de
© Springer-Verlag Berlin Heidelberg 2002
Originally published by Springer-Verlag Berlin Heidelberg New York 2002
Softcover reprint of the hardcover 1st edition 2002

Cover Design: Erich Kirchner
Typesetting: Camera-ready by the editors

SPIN: 11008330   30/3111 – 5 4 3 2 1 – Printed on acid free paper

Preface

When in 1989 the interdisciplinary research program "Stability of Ecosystems", coordinated by the Forest Ecosystems Research Centre of the University of Göttingen, was brought into being, twenty years of experience in forest ecological investigations had already been gathered. As the German part of the International Biological Program (IBP), ecosystem research has been carried out since 1966 in the Solling, an upland region in Northwest Germany. This ecological project, which until now has been the most extensive conducted in Central Europe, was initiated by the vegetation ecologist Heinz Ellenberg and the soil scientist Bernhard Ulrich of the University of Göttingen. In the Solling project scientists from various disciplines were involved, investigating factors which have an effect on primary and secondary production, biodiversity and biogeochemical cycles of beech and spruce forest ecosystems. However, it soon became apparent that this forest area was being influenced by atmospheric pollutants, those had been transported over long distances. Based on long term data on ecosystem ion budgets, the detrimental effects of acid deposition on soils with far-reaching consequences for tree health and ecosystem structure were pointed out by Ulrich in 1979. In the beginning of the 1980s, in different regions of Europe and North America the principal tree species showed obvious symptoms of forest decline. Although the push in forest ecosystem research which was then triggered, could not fully elucidate the causes, there is no doubt that soil acidification and the direct effects of gaseous pollutants strongly reduce the vitality of forest trees. Not only the changes in the element composition in the air and the soil give rise for concern, but also concern was raised about the changes of the climate. The undisputed increase of greenhouse gases, mostly carbon dioxide, is accepted to be the main cause for the global increase in air temperature. However, not the temperature increase of ground near air layers alone, but also the resulting events such as drought and the increase of storm catastrophes render it difficult to predict the development of forests in the future.

The concern about effects of changes in the environment is not limited to the existence and the development of forests. Flooding of entire regions due to the rise of the sea level, endangered food resources, increasing desertification and the direct effects of storm catastrophes on the population are considered to be at least equally important. However, events in the past have shown that forests are more at risk, mostly due to their longevity. The ability of forests to adapt to strong disturbances is flexible compared to other ecosystems.

The functioning of forest ecosystems under complex changes in the environment was the general objective of the second extensive ecosystem research phase in Göttingen. For the coordination of the numerous interdisciplinary research projects in 1989 the Forest Ecosystem Research Centre was founded. Additionally to the Solling, other forest regions were included in the investigations. The research activities of the research centre were combined in two research programs: stability of forest ecosystems and dynamic of changes of forest ecosystems. Each of the research programs was promoted for five years by the financial support of the German Federal Ministry of Research and Technology.

In this volume the research projects of different research realms are portrayed. The studies are not selected according to spatial categories: The investigations expand from forest regions in the North German Pleistocene (chap. 1), beech forests in the Hessian Uplands (chap. 3.2) as well as to spruce and beech stands and re-afforested sites in the Lower Saxony Solling (chaps. 2, 3.1 and 4). The integrative element of all studies is, that they investigate different aspects of the difficult to define construct **forest development.** First of all, forest development can be considered neutrally as a mostly natural forest dynamic. This type of **succession** is exemplarily shown in chapter 1 by a reconstructed time sequence over 300 years of heath landscapes to beech oak climax forest communities.

Along with this mostly undisturbed development of forests, forest development is increasingly viewed in connection with anthropogenic causes for **environmental stress.** The investigations within the roof project in Solling (chap. 2) were aimed at the effects of strong anthropogenic disturbances such as high sulphur and nitrogen inputs on forest ecosystems. Thus, the experimental design was set up to investigate, to what extent a drastic reduction of the element inputs and therefore a relief of the environmental stress induces a stabilisation of the ecosystem equilibrium.

Although the prognoses about the development of our forest ecosystems under environmental stress are not conclusively confirmed, the federal forest administrations in Germany have changed considerably the principles of **forest management** in the past decade. The fundamental principle is an ecologically oriented silviculture, which promotes the natural variety of tree species and natural forest structures. By aiming to increase the diversity of species and structure, it is expected that risk spreading can be achieved and also, that the ability of forest ecosystems to adapt to environmental stress is increased. In order to protect soils and stands, forest technology must be applied carefully. The necessity of liming and melioration measures is still disputed. In chapter 3 ecosystemic relationships in connection with relevant forest management concepts for the future are portrayed. The promotion of natural regeneration processes by irregular opening of the crown cover, as it is required by most forest programs in Central Europe, undoubtedly bears important advantages compared to less natural interventions. However, investigations designed for a medium-term in two beech stands, show that the effects of disturbances on important functions of the ecosystem can not be entirely avoided, even if moderate forms of forest management are applied (chaps. 3.1 and 3.2).

Although the investigations presented in this volume were primarily outlined as basic research, it is still possible to transfer a number of results to applied forest management. This occurs for example, when at problematic sites suitable tree species must be selected and the effect of melioration measures to improve general conditions for new cultivations must be assessed. This is illustrated in chapter 4 by an experiment with newly planted different tree species at a wind-throw site under immission stress. Different, partly extreme interventions to improve the soil condition were compared, and soil chemical changes investigated over several years (chap. 4.1). In addition to the presentation of varying measures suitable for different tree species (chap. 4.2), also the biological influence of agricultural herbs as auxiliary plants on the competitive behaviour of trees is shown (chap. 4.3).

It is apparent that based on the selected contributions compiled in this work, it is not possible to formulate a joint conclusion. The various aspects of forest ecosystem research subsumed here under the construct forest development, emphasize the multi-layered and complex nature of this subject area. Changes in forest ecosystems such as we considered here retrospectively, and at present observe with concern, we can only predict with large uncertainties, as they are arbitrary processes, which can not be explained by linear cause-effect statements. As a consequence, it is logical that possible actions in forestry practise must also be complex, if the objective to respond effectively and contribute to a stabilisation of the system, is to be achieved.

Göttingen, July 2002   Achim Dohrenbusch and Norbert Bartsch

Contents

Contributors

Bartsch, N.

Institute of Silviculture, Göttingen University, Büsgenweg 1,
D-37077 Göttingen, Germany, e-mail: n.bartsch@forst.uni-goettingen.de

Bauhus, J.

School of Resources, Environment & Society, The Australian National
University, Canberra, ACT 0200, Australia,
e-mail: juergen.bauhus@anu.edu.au

Bredemeier, M.

Forest Ecosystems Research Center, Büsgenweg 1,
D-37077 Göttingen, Germany, e-mail: mbredem@gwdg.de

Dohrenbusch, A.

Institute of Silviculture, Göttingen University, Büsgenweg 1,
D-37077 Göttingen, Germany, e-mail: adohren@gwdg.de

Fritz, H. W.

Institute of Forest Botany, Büsgenweg 2, D-37077 Göttingen, Germany

Godt, J.

Faculty of Urban and Landscape Planning, Department Landscape
Ecology and Soil Science, Kassel University, Gottschalkstr. 28,
D-34109 Kassel, Germany, e-mail: jgodt@hrz.uni-kassel.de

Jaehne, S.

Institute of Silviculture, Göttingen University, Büsgenweg 1,
D-37077 Göttingen, Germany, e-mail: ufwb@gwdg.de

Kumke, J.

Institute of Silviculture, Göttingen University, Büsgenweg 1,
D-37077 Göttingen, Germany, e-mail: ufwb@gwdg.de

Lamersdorf, N.

Institute of Soil Science and Forest Nutrition, Büsgenweg 2,
D-37077 Göttingen, Germany, e-mail: nlamers@gwdg.de

Leuschner, C.

Plant Ecology, Albrecht-von-Haller-Institute of Plant Sciences,
Göttingen University, Untere Karspüle 2, D-37073 Göttingen, Germany,
e-mail: cleusch@gwdg.de

Linke, J.

Institute of Forest Botany and Forest Zoology, Dresden University of
Technology, Pienner Straße 7, D-01735 Tharandt, Germany,
e-mail: forstbot@forst.tu-dresden.de

Mackenthun, G.

Institute of Silviculture, Göttingen University, Büsgenweg 1,
D-37077 Göttingen, Germany, e-mail: ufwb@gwdg.de

Meiwes, K. J.

Forest Reseach Institute of Lower Saxony, Grätzelstraße 2,
D-37079 Göttingen, Germany, e-mail: meiwes@nfv.gwdg.de

Meyer, A. C.

Institute of Silviculture, Göttingen University, Büsgenweg 1,
D-37077 Göttingen, Germany, e-mail: ufwb@gwdg.de

Roloff, A.

Institute of Forest Botany and Forest Zoology, Dresden University of Technology, Pienner Straße 7, D-01735 Tharandt, Germany, e-mail: forstbot@forst.tu-dresden.de

Vor, T.

Institute of Silviculture, Göttingen University, Büsgenweg 1, D-37077 Göttingen, Germany, e-mail: tvor@gwdg.de

1 Forest succession and water resources: soil hydrology and ecosystem water turnover in early, mid and late stages of a 300-yr-long chronosequence on sandy soil

C. Leuschner

Plant Ecology, Albrecht-von-Haller-Institute of Plant Sciences,
Göttingen University, Untere Karspüle 2, D-37073 Göttingen, Germany,
e-mail: cleusch@gwdg.de

Abstract

Information on changes in water availability and water turnover in the course of long-term secondary successions is very limited. This study used a chronosequence approach to quantify the soil moisture status (water content and matric potentials) and the water fluxes (bulk precipitation, throughfall, interception, stemflow, transpiration, evaporation, drainage) in three stages of a 300-yr-long succession from *Calluna vulgaris* heathland to *Fagus sylvatica-Quercus petraea* late-successional forest (heathland-to-forest succession) on sandy soil in Northwest Germany. Evapotranspiration losses increased early in succession as the vegetation gained in height and developed larger leaf areas, but they tended to decrease toward the late-successional community again. In the early-to-mid successional community (*Betula pendula-Pinus sylvestris* pioneer forest), water uptake (or transpiration loss) peaked and drainage losses reached their minimum during the succession. These changes in the water fluxes reflect two underlying processes, (a) the shift from drainage to evapotranspiration losses and, (b) the relative change from interception to transpiration with proceeding succession. It is concluded that the large increase in evapotranspiration, that follows the establishment of trees in this succession, ultimately is driven by three processes, (i) an increasing canopy-atmosphere coupling due to a more elevated and rougher vegetation surface which facilitates gas exchange, (ii) a threefold increase in the stand leaf area which increases the transpiring and intercepting surface, and (iii) a deeper penetration of the root systems with proceeding succession that allows access to larger soil water pools.

The physical analysis of the sandy mineral soils revealed some characteristics that affect soil moisture availability during this succession. The size of the soil water reserves was remarkably similar among the three communities despite significant differences in interception and transpiration rates. Profiles under all three communities showed an anomaly in their water content and matric potential patterns with minima occurring at 40-60 cm depth rather than at the soil surface with highest fine root densities. Compared to literature data, the sandy mineral soils under the three successional communities were characterised by remarkably high field capacities (150 to 170 mm of plant-available water per 1 m profile) which results from the fact that water held at potentials of -20 to -100 hPa (conventionally assessed as percolating water) contributed to the plant-available water pool as well. Additional soil water storage resulted from the accumulation of organic material on the forest floor during succession which significantly increased the soil water reserves in late stages. This process represents a case of facilitation since water availability is improved by autogenic processes.

Key words: drainage, evapotranspiration, heathland, interception, late-successional forest, matric potential, NW Germany, pioneer forest, secondary forest succession, soil water content

1.1 Introduction

Vegetation structure is a key attribute that can exert large influences on the fluxes of radiation, water and nutrients in ecosystems. For example, transpiration, canopy interception and stemflow are highly dependent on vegetation height, plant surface area, and the arrangement of above-ground plant organs in the canopy space (Calder 1990). Vegetation change, which may result from natural plant replacement processes or human disturbance, is likely to change ecosystem water turnover and soil water availability. Many forest hydrologists have studied the marked alterations in runoff, drainage, evapotranspiration and soil moisture content which result from the conversion of forest to non-woody vegetation (Greenwood 1992). Similarly, agricultural scientists have devoted much attention to the relationship between crop species or culture techniques and evapotranspiration in regions with water shortage (Stewart and Nielsen 1990). However, much less is known about changes in water turnover that result from long-term plant succession, i.e. the natural replacement of herbaceous or dwarf-shrub communities by forest. The natural re-establishment of forest is an important process in non-permanent landuse systems (such as slash-and-burn techniques or rotation systems with long fallow phases) and in marginal regions where agriculture has been abandoned permanently due to economic reasons.

On nutrient-poor soils in Central Europe, agriculture was either intensified, or, in the case of low-productive farming systems, was given up with the introduction of industrial fertilisers. During the past 200 years, large areas formerly used as grazed heathlands, arable fields or pastures were abandoned in the sandy diluvial regions of Northern Germany, Denmark, the Netherlands and Belgium. This initiated two processes, (i) the planting of conifers (mainly of *Pinus sylvestris* L. and *Picea abies* Karst.) and (ii) secondary succession in the direction of broad-leaved forest (Leuschner 1994, 2001), that changed the character and the hydrology of the landscape fundamentally. In the Lüneburger Heide region of NW Germany, sheep grazing on *Calluna vulgaris* L. heathland was widely abandoned from about 1780 onwards and vast areas were left for reforestation with *P. sylvestris,* or underwent natural secondary succession back towards pioneer and late-successional forest communities (Griese 1987, Leuschner 1994)

Succession research has been a primarily descriptive discipline for long time with most studies focussing on species composition and vegetation structure (Miles 1979). The causes of vegetation change were studied in a number of primary and secondary successions with an emphasis on plant resources (e.g. Viereck 1970; Vitousek and Reiners 1975; Christensen and Peet 1981; Walker *et al.* 1981; Tilman 1988; Bormann and Sidle 1990; Leuschner et al. 1993; Gerlach *et al.* 1994; Emmer 1995). Indeed, plant resources such as light, water and nutrients are viewed as key factors in succession (Bazzaz and Sipe 1987; Leuschner and Rode 1999). However, most of the investigations on the mechanisms of succession have concentrated on light and nitrogen availability only. Similarly, Tilman's (1985) formulation of the resource ratio hypothesis focuses on two plant resources, light availability at the soil surface and nitrogen concentration in the soil. Remarkably, water as an important plant resource has been widely ignored in succession research.

Causal explanations of vegetation change should base on an understanding of resource dynamics rather than on static descriptions of resource availability in successional environments. Information on resource fluxes and ecosystem resource budgets in the context of succession is much more limited than are data on successional changes in resource concentrations. Among the few examples of flux studies are the investigations of Berendse (1990), Kellman and Roulet (1990), Rode et al. (1993), Gerlach et al. (1994) and Rode (1995) that consider ecosystem nutrient turnover in successional communities. For water as a resource, detailed studies on changes in availability and ecosystem turnover during long-term forest succession are still lacking. Such information, however, is needed when the water yield of landscapes that undergo successional change is to be predicted. Moreover, a better understanding of the causes, that underly species turnover during succession, can be expected since successional plants often compete for water as a limiting resource.

This paper summarises the results of hydrological studies conducted in the years 1990 to 1992 in a chronosequence spanning the succession from heathland to deciduous oak-beech forest in the Lüneburger Heide region of NW Germany. Principal pathways of this secondary succession are described in Leuschner (1994). Three stages of the about 300-ys-long sequence were selected for comparative study, (i) the intitial heathland (*Calluna vulgaris)* stage, (ii) the birch-pine pioneer forest (*Betula pendula* Roth, *Pinus sylvestris*) stage, and (iii) the oak-beech late-successional forest (*Quercus petraea* (Matt.) Liebl., *Fagus sylvatica* L.) stage. The investigations focussed on (i) soil water availability (water reserves) and ecosystem water turnover (water fluxes) by using methods from soil physics, forest hydrology and microclimatology. The aims of the study were (i) to characterise the soil water availability in early, mid and late stages of the heathland-to-forest succession in its temporal and spatial variability, (ii) to quantify the water turnover (input, storage, losses) in three important stages of the succession, (iii) to analyse succession-related trends in water availability and water fluxes, and (iv) to identify structural attributes of species and communities of the succession that determine how water is channeled through the ecosystem. This review covers only water. Results on fluxes of radiation, carbon and nutrients in the successional communities are presented in Leuschner and Rode (1999) and Leuschner (2002), data on the ecophysiology of the dominant plant species of the succession are given in Rode and Schmitt (1995) and Leuschner (2001).

1.2 Methods

1.2.1 Study sites

The study was carried out in the Northwest German state of Lower Saxony in the surroundings of Unterlüß (southeastern part of the Lüneburger Heide, 52°45'N, 10°30'E). Two examples each of the initial stage (heathland), the early-to-mid stage (birch-pine pioneer forest) and the late stage (oak-beech forest) of a heathland-to-forest succession on comparable geological parent material were chosen for study (Table 1-1; sites CH1 and BP3 at Rahberg, 7 km E of Munster; CH2 and BP4 at Schillohsberg, 4 km W of Unterlüß; OB5 2 km W of Unterlüß; OB6 2 km E of Unterlüß).

Table 1-1. Some characteristics of the six research sites.

Stage in succession	Plot	Vegetation type	Community	Dominant plant species[a]	Vegetation height (max.) (m)	Max. age of plants (years)	Approx. age in heathland-forest succ. (years)	Humus accumulation period (years)[b]	Soil type, humus form[c]
Initial	CH1	Heathland with *Pinus/ Betula* colonisation	Genisto-Callunetum Cladonietosum	*Calluna vulg.* (*Pinus sylv., Betula pend.*)	0.3 (1.0)	28 (*Pinus*: 5)	5	>10	Orthic Podzol Xeromor
Initial	CH2	Heathland with *Pinus/ Betula* colonisation	Genisto-Callunetum cladonietosum	*Calluna vulg.* (*Pinus sylv., Betula pend.*)	0.3 (1.0)	30-35 (*Pinus*: 10)	10	>10	Orthic Podzol Xeromor
Early/mid	BP3	Pioneer birch-pine forest	Leucobryo-Pinetum	*Pinus sylv.* (73%) *Betula pend.* (27%) H: *Avenella flex.*	13	35	ca. 35	ca. 85	Orthic Podzol[f] Hemimor
Early/mid	BP4	Pioneer birch-pine forest	Cladonio/ Leucobryo-Pinetum	*Pinus sylv.* (97%) *Betula pend.* (3%) H: *Vaccinium* spp.	16	44	ca. 45	ca. 45	Orthic Podzol[f] Hemimor
Late[e]	OB5	Oak-beech forest	Luzulo-Fagetum (lowland type)[d]	*Fagus sylv.* (77%) *Quercus petr.* (23%)	32	110/200	>300	>300	Spodo-dystric Cambisol Hemihumimor/ Hemimor
Late[e]	OB6	Oak-beech forest	Luzulo-Fagetum (lowland type)[d]	*Fagus sylv.* (55%) *Quercus petr.* (45%)	34	175	>300	>300	Spodo-dystric Cambisol Hemihumimor/ Hemimor

[a]Values in percent for the dominant plant species refer to the number of stems in the canopy layer. [b]Humus accumulation since last major disturbance (sodcutting, burning). [c]Humus form according to Klinka et al. (1993). [d]Synonymous to Deschampsio Fagetum. [e]Putative terminal stage of succession. [f]Podzol with several properties of Spodo-dystric Cambisols (see Leuschner et al. 1993). H, herbaceous layer.

In a phytosociological study more than 200 plant sociological samples were done in forest stands in the region (Heinken 1995). Plot selection criteria were (i) vegetation composition, (ii) site history, (iii) comparability of geological substrate, (iv) intensity of human impact, and (v) vicinity to other stages. The maximum distance between the six selected sites is about 18 km. The oak-beech forest community is assumed to represent the autochthonous (potential natural) forest vegetation on this substrate. In the two chosen stands of this community (OB5 and OB6), litter removal and use as wood-pasture most likely have occurred in the past, but a conversion into heathland might not have happened. Most hydrological studies, in particular the flux measurements, concentrated on the sites CH1, BP3 and OB5.

The region of Unterlüß is part of the diluvial lowlands of NW Germany that are shaped by the deposits of the penultimate (Saale) Ice Age which later were influenced by periglacial processes of the last (Weichsel) Ice Age. The substrate mainly consists of sandy, nutrient-poor sediments of the Drenthe-2-stadial of the Saale Ice Age. The lower part of all investigated soil profiles (nine per site) consists of stratified fluvio-glacial melt-water sand (locally also sandy basal moraine material). Cover sand of periglacial origin forms the upper 20 (to 60) cm of the profiles. Medium-grained sand is the dominant size fraction of the sediments (40-60% of dry mass) while the clay content is very low (typically <5%). Cation exchange capacity and 'base saturation' are also very low in the profiles under the three successional communities (Leuschner and Rode 1999). The heathland soils are topped by dry Xeromors, the birch-pine and oak-beech forest soils bear thick Hemihumimors of 8-10 cm depth. All sites are on level terrain at 90-110 m above sea level. In all three communities, fine roots are highly concentrated in the topsoil and fine root densities decrease exponentially with soil depth (Leuschner 2001). The ground water table is found more than 50 m below the rooting zone. No soil liming has been conducted. The climate is of a humid suboceanic type (annual mean air temperature: 8.1 °C, annual precipitation: 801 mm). Atmospheric deposition of nitrogen at the study sites is comparably low for Central European conditions (Rode 1999). Until recently, no marked changes in vegetation structure due to eutrophication (e.g. invasion of grasses into heathlands) have occurred.

1.2.2 Hydrological methods

Soil water status

Soil water contents (θ) and soil matric potentials (Ψ_m) were measured separately in the mineral soil and the organic layers (forest floor) of the three communities. The *water content* of the mineral soil was investigated gravimetrically at 100 cm^3 samples taken weekly in 0-5, 10-15 and 55-60 cm depth (3 to 5 replicates each per community). The water content of the organic layers was determined gravimetrically at samples of the O_f- and O_h-layers. Plots under *Betula* or *Pinus* trees (site

BP3) and *Quercus* or *Fagus* trees (OB5) were sampled separately due to species-specific differences in humus structure. From May 1991 to December 1992, eight samples each per tree species were taken weekly (in summer) or 2-4 weekly (in winter) with a 5 cm-diameter root corer at 40 to 200 cm distance from a stem. By simultaneous measurement of the profile depth in undisturbed samples, the water content data could be expressed as volumetric water content (vol. %) and in terms of water storage (mm per profile). Continuous measurements of the *soil matric potential* were conducted in the vegetation periods (April to October) of 1990, 1991 and 1992 in the pioneer and the late-successional forest communities with tensiometers that were arranged in stem-centred schemes around diameter-representative *Betula* and *Pinus* (site BP3) or *Fagus* and *Quercus* stems (OB5). Depths of 5, 15, 45, 80, 130 and 170 cm, and stem distances of 50, 100, 150, 200, 250, 300 and 350 cm were instrumented (Fig. 1-1). Two arrays each per tree species were installed and recording took place every 3 to 7 days during the seasons. In the heathland (CH1), 10 tensiometers each were installed at 15 and 60 cm depth. During periods of drought, additional soil water potential measurements were irregularly conducted in the organic layers and the mineral soil with PCT-55 soil psychrometers (Wescor, Logan, Utah, USA).

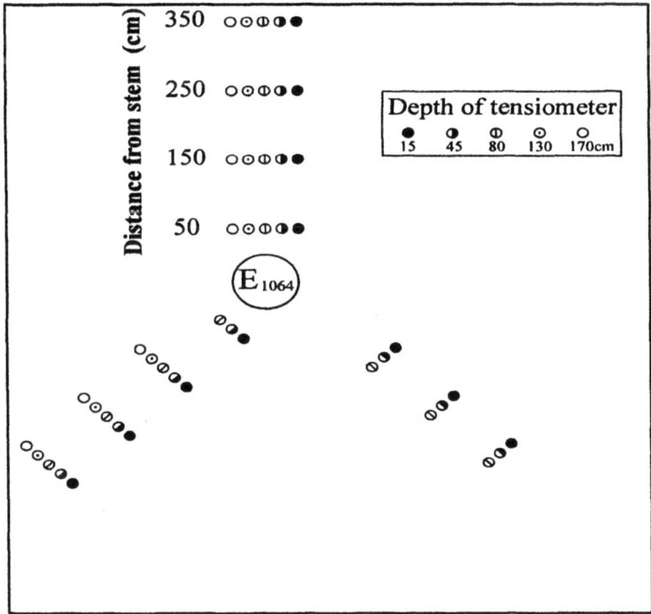

Fig. 1-1. Spatial arrangement of tensiometers in the vicinity of a tree (here: oak no. 1064). Six depths in the mineral soil (5 - 170 cm) and four distances to the stem (0.5 to 3.5 m) were instrumented.

Water retention curves (i.e. the relationship between soil water matric potential and volumetric water content, θ/Ψ_m curves) were measured in the laboratory at 'undisturbed' samples by desorption with hanging water columns (Leuschner 1998). In the mineral soil, samples of 100 cm^3 volume taken at four mineral soil depths were investigated (5 replicates). Each five samples of *Quercus, Fagus, Betula* and *Pinus* humus (250 cm^3 volume) were also analysed. Synchronous measurements with tensiometers and tdr-probes were used to establish θ/Ψ_m *relationships in the field.* Water held at matric potentials < -15000 hPa was termed 'non-root-extractable', water held between -2.5 and -15000 hPa was considered as 'plant-available'. The water content immediately after a saturating infiltration was taken as the 'saturated water content' θ_s. This is less than the maximum water content θ_{max} (= porosity) with all air space filled with water. The *relationship between water content and hydraulic conductivity* was determined in the laboratory for mineral soil samples (250 cm^3) by desorption with hanging water columns (3-4 soil depths, 5 replicates each). Laboratory infiltration experiments were also conducted with humus (O_{fh}) material of the oak-beech and birch-pine forests to establish empirical *relationships between rainfall amount, water retention (wetting curves) and resulting percolation loss.* Undisturbed forest floor sods of 17 x 37 cm size (sampled under *Fagus* or *Pinus*) were treated with artificial rain of 0.5 to 30 mm height. The sod weight was determined 5 min after application and the retained and the percolated water were expressed as a function of rainfall and initial humus water content. This procedure was repeated with sods of variable moisture content (10 to 31.5 mm of initial water storage). Each treatment consisted of 5 replicate measurements that were averaged (Leuschner 1998).

Water fluxes

Above-ground water fluxes (precipitation, throughfall, stemflow, interception, soil evaporation, transpiration) were measured at the sites CH1, BP3 and OB5 with standard methods as described in Table 1-2. Soil water content data were used to calculate the seasonal change in *water storage in the soil profiles* downward to 70 cm depth (significant fine root densities are found up to 50-70 cm depth; Hertel 1999). Patch sizes large enough to fulfil the fetch requirements of micrometeorological gradient measurements of *evapotranspiration* existed only in the initial heathland community (site CH1) and in the late-successional oak-beech forest (OB5). Using scaffolding towers at these two sites, the vertical fluxes of latent and sensible heat above the canopies were calculated according to the Bowen ratio energy balance approach. It is based on continuous high precision temperature and humidity gradient measurements (Baldocchi et al. 1988), records of net radiation, and calculations of heat fluxes into the soil and biomass.

Table. 1-2. Overview of the methods used to measure soil water status and ecosystem water fluxes at the study sites CH1 (heathland), BP3 (pioneer birch-pine forest) and OB5 (late-successional oak-beech forest) (Sym, Symbol; L, level of study, i.e. plant (P) or ecosystem (E); n, number of replicates).

Variable	Sym	Site	L	Methods and instrumentation	n	Explanations
Bulk precipitation	BP	CH1	E	wind-shielded PE gauges (314 cm^2)	4	at 1.2 m height above heathland, sampled 2-weekly
		BP3	E	(data of adjacent CH1 site used)		
		OB5	E	wind-shielded PE gauges (314 cm^2)	4	at canopy height (30 m), sampled 2-weekly
Throughfall	TF	CH1	E	50 cm^2 PVC-grooves below the heath canopy	18	sampled 2-weekly
		BP3,OB5	E	systematically positioned rain gauges	36	sampled 2-weekly
Stemflow	SF	CH1	E	not measured, guessed to be 5% of BP		
		BP3	P	Collars at each 4 pine and birch trees	8	sampled 2-weekly
		OB5	P	Collars at 13 beech and 3 oak trees	16	all trees in a 500 m^2 area included, sampled 2-weekly
Canopy interception	IN	CH1,BP3, OB5	E	IN = BP - TF - SF		
Stand evapo-transpiration	ET	CH1	E	Bowen ratio/energy balance approach	1	based on dT/dz, de/dz, Q_n, soil heat flux; 2 m tower
		BP3	E	Penman-Monteith equat. (big leaf formul.)	1	$r_c + r_a$ values of Fuhrberg pineforest used, Flüggen 1991
		OB5	E	Bowen ratio/energy balance approach	1	based on dT/dz, de/dz, Q_n + soil heat flux; 36m tower
Soil evaporation	E_s	CH1	E	not determined		not distinguished from ET
		BP3,OB5	E	Simplify. energy bal. calc. (Leuschner 1997)	1	mainly based on Q_n, T and Δe data at soil surface
Drainage	D	CH1,BP3, OB5	E	annual basis: D = BP - ET - IN - ΔTW	1	ΔTW = change in soil water content
		OB5	E	daily basis: one-dim. soil water flux model	1	Darcy equation, tensiom. data at 15, 60, 120 cm depth
Root water extraction	E	OB5	E	E = ET - E_s	1	

In the case of the birch-pine pioneer forest community, evapotranspiration was estimated by the 'big leaf' formulation of the Penman-Monteith equation with net radiation, temperature and air humidity measured above the canopy (Brutsaert 1982). In this approach, values of canopy (r_c) and aerodynamic resistances (r_a) measured by Flüggen (1991) at the Fuhrberg pine forest (Lüneburger Heide, 40 km south of BP3) were used.

For estimating drainage losses under the heathland, pioneer forest and late-successional forest communities, simple *water budget calculations* at the ecosystem level were conducted for the year 1991. They base on annual totals of precipitation and stand evapotranspiration, and data on changes in soil water storage. Water flux calculations with a one-dimensional model based on tensiometer data and the Darcy equation were also available for estimating drainage in the late-successional forest (Hölscher and Leuschner, unpublished).

For quantifying the *water turnover of the organic layers*, an attempt was made to measure the relevant water fluxes with appropriate techniques directly in the field and to describe the water turnover with a one-dimensional model ('Forest Floor Water Flux Model') in temporal resolution of one day. Details on these measurements and the model structure are given in Leuschner (1998). Briefly, water input into the humus equals canopy throughfall (TF) plus stemflow (SF); the model, however, neglects stemflow due to its non-homogenous distribution at the forest floor. Output terms are (i) the percolation out of the organic profile into the mineral topsoil (seepage, SP), (ii) evaporation from the litter surface (EV), (iii) flux into or out of the storage in the profile (ST), and (iv) water uptake by fine roots in the densely rooted organic horizons (UP). Capillary rise from the mineral soil is neglected. The model uses a mass balance approach and bases on empirically established relationships between rainfall amount, water retention of the humus material (wetting curves) and resulting percolation loss (see above). As input data the model requires daily throughfall and stand microclimatological data, and the humus moisture content at weekly intervals. After solving the water balance equation, the resulting term is taken as the water uptake by roots in the organic profile (UP):

$$(1) \quad TF + EV + SP + ST + UP = 0$$

For assessing the relative contribution of root water uptake from the organic layers or the mineral soil horizons, the uptake rates as calculated with the Forest Floor Water Flux Model were related to evapotranspiration (ET) data at the stand level (Bowen ratio energy balance calculations or Penman-Monteith calculations for the oak-beech forest and birch-pine forest, respectively). On rainless days, the calculated root water uptake rate in the organic layers (UP) and the litter evaporation rate (EV) both were subtracted from ET to estimate the water extraction by roots in the mineral soil profile alone. This procedure allowed a rough estimation of the relative importance of water uptake by the tree roots in the organic layers.

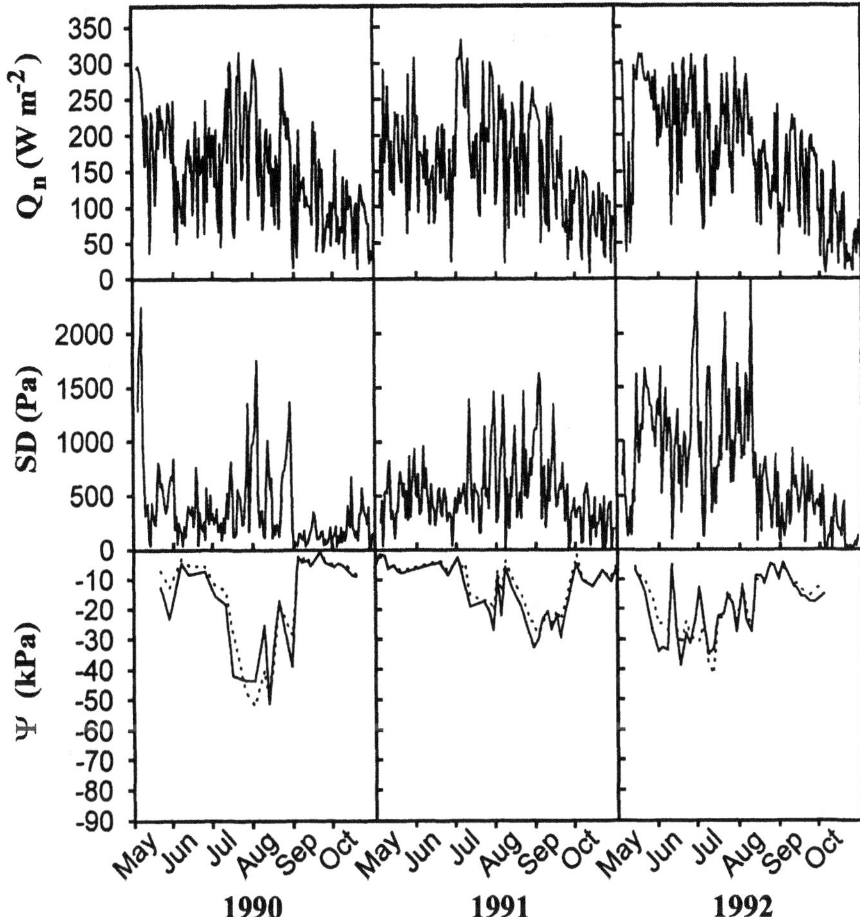

Fig. 1-2. Seasonal courses of net radiation input at canopy height (Q_n, daily totals between dawn and dusk), vapor pressure saturation deficit of the air at sun canopy height (SD, daily averages between dawn and dusk), and soil matric potentials at 15 cm depth (Ψ, measured separately in the vicinity of *Quercus* (solid line) and *Fagus* stems (dotted line) with each 22 tensiometers) during the growing seasons 1990 to 1992 at the oak-beech forest site.

Hydrologic regimes in the three studied summers

The summers of 1990 to 1992 differed markedly with respect to radiation input, atmospheric water demand and soil water availability (Fig. 1-2 and Table 1-3). In *1990*, a long and physiologically important drought period occurred in July/August at the peak of physiological activity. During this drought, very low tensiometer readings were recorded at the three sites. The summer *1991* was characterised by frost damage in May/June and a severe drought late in summer (August/September). The summer *1992* was exceptionally hot with high radiation loads and a high atmospheric demand; moderate drought periods occurred early in summer (June/July) but not during mid-summer. Figure 1-3 contrasts the monthly rainfall amounts in the three-year-period with the thirty-year-mean. More details on the climatic variability in the study period are given in Backes and Leuschner (2000).

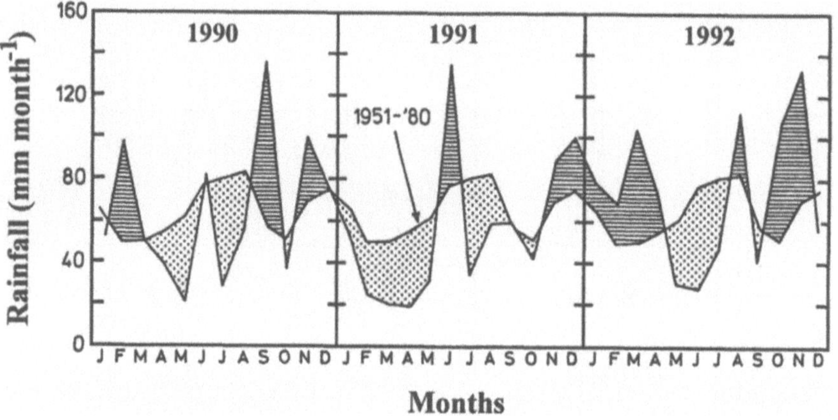

Fig. 1-3. Monthly rainfall amounts in 1990-1992 and thirty-year average at the meteorological station Unterlüß (Deutscher Wetterdienst, Offenbach).

Table 1-3. Some meteorological and soil hydrological parameters that characterise the variability of soil drought intensity in the summers 1990-1992 (data of the site OB5; Ψ_m - seasonal minimum of the soil matric potential at 15 cm depth as measured with tensiometers, n = 44; θ - seasonal minimum of the volumetric water content of the soil at 15 cm depth as determined gravimetrically, n = 3-5; rainfall, air temperature and incoming solar radiation refer to the period May 1 to September 30; solar radiation - mean daily fluxes).

	Ψ_m (kPa)	θ (vol. %)	Rainfall (mm)	Temperature (°C)	Solar radiation (MJ m⁻² d⁻¹)
1990	-51	2	320	14.4	15.0
1991	-31	5	320	13.5	15.3
1992	-40	5	259	15.3	17.2

Statistical analysis

A non-parametric Mann-Whitney (Wilcoxon) two-sample test (5% rejection level) was used to test for significant differences in hydrologic parameters between any of the three communities. For a number of flux parameters such as evapotranspiration, drainage loss or canopy interception, no replicate data at the ecosystem level could be obtained and, thus, no statistical treatment was conducted. The underlying data, however, were sampled at sufficient replication (see Table 1-2).

1.3 Water resources during forest succession

In order to predict changes in hydrology as a consequence of long-term secondary and primary succession, the relationship between vegetation structure and water turnover has to be analysed. Succession may influence the plant availability of water and the flux of water through ecosystems by six processes, three of which are 'non-uptake effects' on resources and three are 'uptake effects' in the sense of Goldberg (1990). They are given below.

(A) Uptake effects

The influence of uptake effects on water availability and water fluxes grows with increasing plant water uptake or transpiration.

1) Plant water loss through transpiration is partly under stomatal and leaf area control and also depends on vegetation height (plant morphology and physiological activity). The amount of transpiration simultaneously depends on, and has an influence on, the availability of soil water.

2) Annual transpiration and interception losses of a given vegetation type are also dependent on the length of the leafy period (plant phenology).

3) Plant-specific patterns of vertical root extension may determine the amount of water that is available to a plant; deep rooting may be a prerequisite of a higher transpiration rate (plant morphology and physiological activity).

(B) Non-uptake effects

Plants can influence water availability or water fluxes by processes other than water uptake and transpiration. Therefore, the size of these effects is not dependent on the rate of resource (water) consumption (Leuschner 1993b).

4) Interception loss is an important component of evapotranspiration. It depends on the size and structure of plant surfaces and influences the amount of water that infiltrates into the soil. Thus, it is important for soil water availability.

5) Plant productivity and chemical composition of organic debris influence the thickness of the organic layers that accumulate at the forest floor. Organic layers can play an important role as an additional soil water pool.

6) Plant canopy structure and leaf area index control the net radiation flux density at the forest floor, which essentially determines soil evaporation and, thus, soil water availability.

The recent study was guided by the following four questions concerning successional changes in water availability and water turnover:

(i) Do transpiration, interception and evaporation (i.e. the three water vapour fluxes) change in relation to each other during succession?

(ii) Does a relationship exist between water uptake and above-ground (leaf) and below-ground (fine root) surfaces that allows to predict plant water use during succession?

(iii) What changes occur with respect to drainage?

(iv) What changes occur with respect to soil water storage? Is soil water availability negatively related to water uptake?

1.4 Changes in soil hydrology

1.4.1 Soil physical properties

1.4.1.1 Mineral soil

Medium-grained sand is the dominant size class of the sediments that comprises typically 40-60% of the solid matter in the profiles under the three successional communities (Fig. 1-4). The content of finer particles (silt and clay) with higher matrical forces is very low. Although a considerable horizontal and vertical variability in sediment composition exists, clay contents typically range between 1% and 5% in the upper mineral soil profiles (0-40 cm, mainly periglacial cover sand) and decrease to 0-2% in the lower profiles (40-130 cm, fluvioglacial melt water sand) with only minor differences between heathland, birch-pine and oak-beech forest communities (Fig. 1-5).

The density per soil volume of particles < 2 mm increases in all profiles from 1.0-1.3 g cm^{-3} at the soil surface to 1.4-1.5 g cm^{-3} at 1 m depth. Highly porous gravel layers with particles > 2 mm that occupy more than 20% of the volume occur locally in the profiles; they facilitate downward water movement in the sandy profiles considerably. Profiles under the oak-beech stands (sites OB5 and OB6) have significantly smaller dry mass densities in their topsoils (1.0 to 1.1 g cm^{-3}) than profiles under heathland and birch-pine forests (1.1 to 1.3 g cm^{-3}). This differ-ence is thought to result from a higher density of coarse and large roots in the oak-beech forest community, which counteracts soil compaction. Differences in clay or organic matter content, which may also influence soil density, are of minor importance.

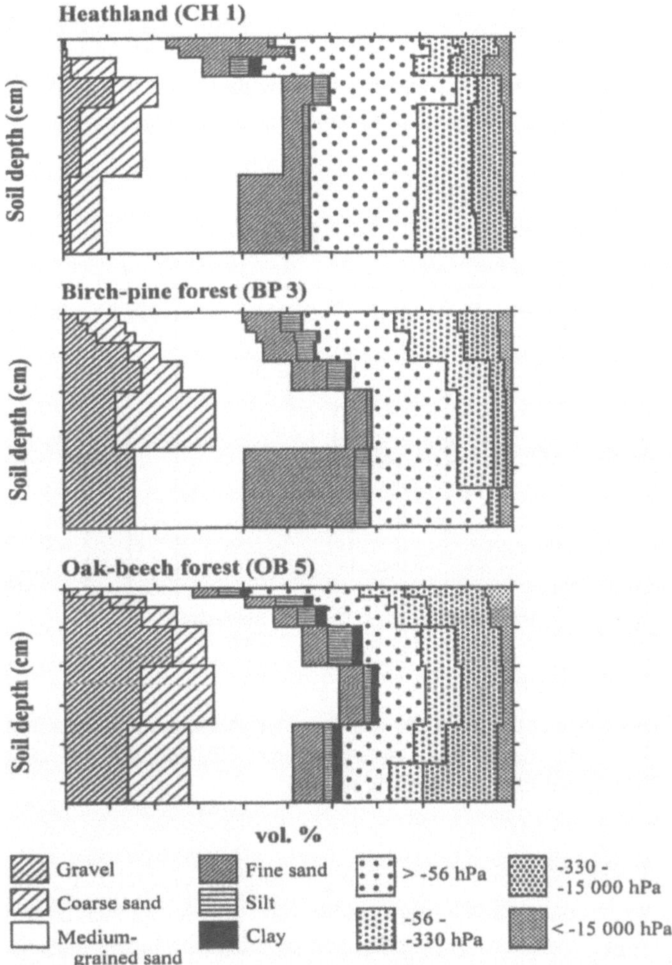

Fig. 1-4. Grain size fractions and pore size distribution in soil profiles under heathland (site CH1), birch-pine forest (BP3) and oak-beech forest (OB5, in percent of the total soil volume). Gravel (> 2 mm), coarse sand (630-2000 µm), medium-grained sand (200-630 µm), fine sand (63 - 200 µm), silt (2-63 µm), clay (< 2 µm), large pores (referring to matric potentials of -330 to -56 hPa and > -56 hPa), medium-sized pores (-15000 to -330 hPa), fine pores (< -15000 hPa). Four replicate samples each were analysed in the laboratory by desorption.

The water content/matric potential relationship (θ/Ψ_m curve) shall be analysed from the results of laboratory studies (desorption by suction) and field measurements (synchronous operation of tdr probes and tensiometers). All soil samples of the three communities investigated have porosites (or maximum water contents θ_{max}) of 36 to 42% (Fig. 1-6). The laboratory θ/Ψ_m curves are similar for the profiles under the three communities and show a rapid loss of water in the -10 to -100 hPa (log 1 to 2) potential range due to the dominance of large pores in the sandy material (compare Fig. 1-4). An exception exists in the case of the mineral topsoil of the oak-beech forest (OB5) that has a particularly low dry mass density and a corresponding high θ_{max} value of 46 to 55% (Fig. 1-6: bottom panel).

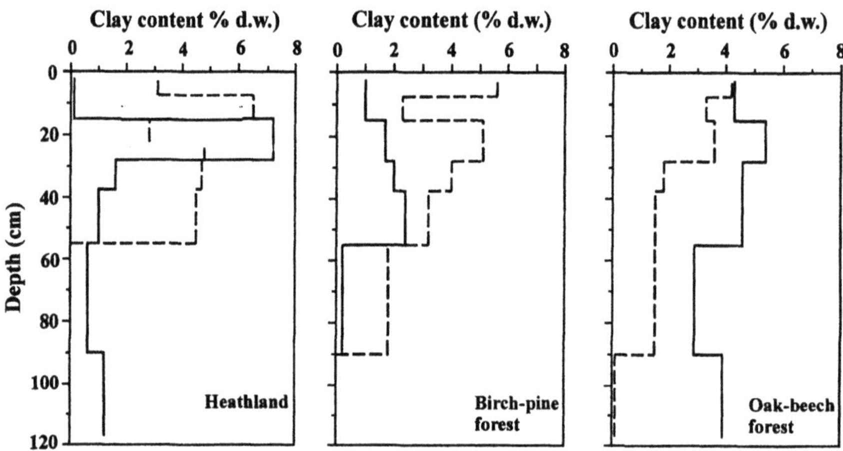

Fig. 1-5. Clay content of the soil (in % of soil dry weight) in profiles under heathland (sites CH1 - bold line; CH2 - dashed line), birch-pine forest (BP3 and BP4) and oak-beech forest (OB5 and OB6). Means of 4 to 17 profiles per site.

This soil material shows a contrasting desorption characteristic with a higher maximum water content and a more gentle water loss for a potential drop between -10 and -100 hPa compared to the other soil samples with higher densities. The fact that soil water contents were particularly high in this horizon during the study period is seen as a consequence of this high porosity value.

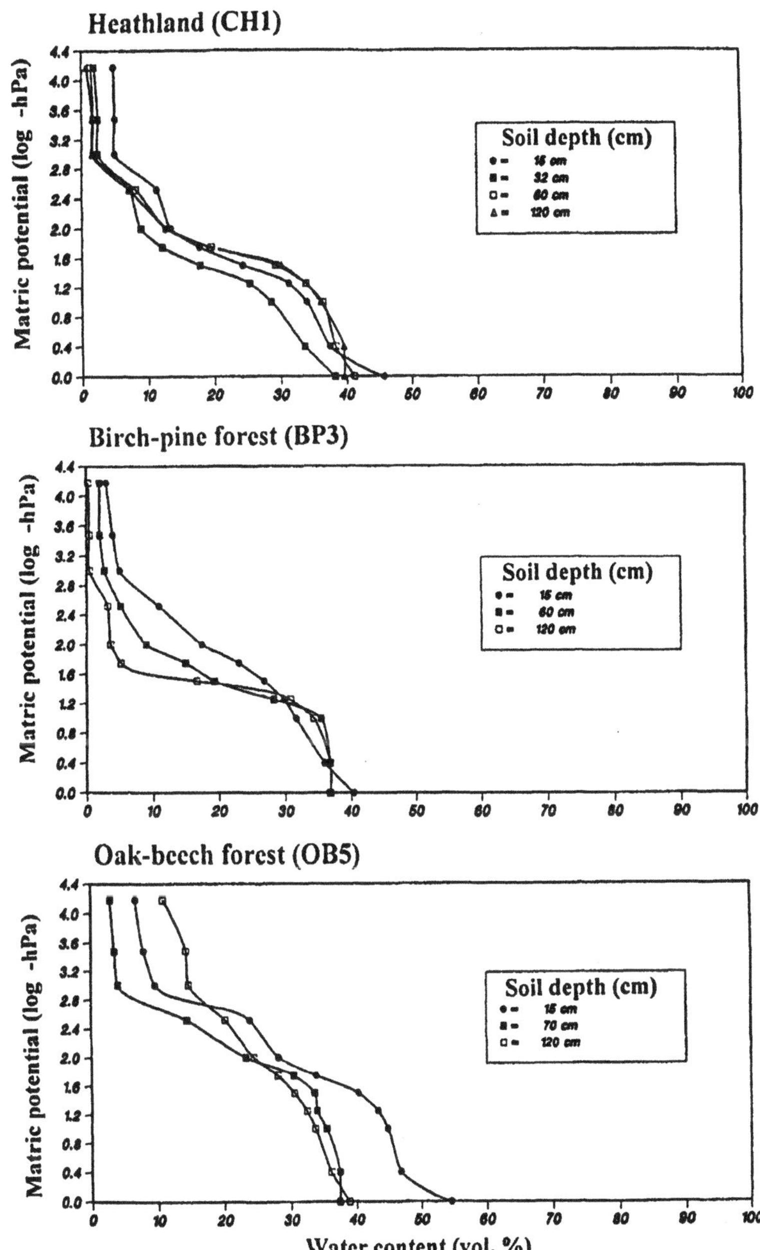

Fig. 1-6. Relationship between volumetric soil water content and soil matric potential at three or four soil depths under heathland, birch-pine forest and oak-beech forest as determined by desorption in the laboratory (mean of five replicates each).

Synchronous measurements of θ and Ψ_m by tdr probes and tensiometers during various drying and rewetting periods of 1992 or 1995 in the soil under the birch-pine forest (Fig. 1-7) and the oak-beech forest (Fig. 1-8) show that laboratory and field θ/Ψ_m curves may differ considerably with field-derived curves being shifted to lower water contents (Fig. 1-7). Moreover, large differences in θ/Ψ_m relationships seem to exist among different soil depths of a single soil profile. Horizons with lower soil densities (e.g. the 5 cm depth in Fig. 1-8) may have water contents at a given matric potential that are twofold higher than those of subsoil horizons with higher densities.

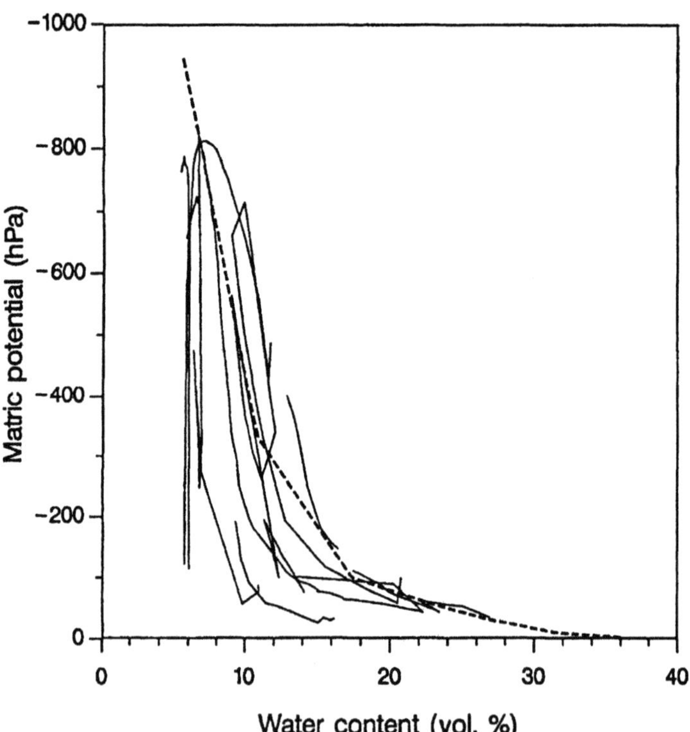

Fig. 1-7. Relationship between volumetric water content at 15 cm depth (tdr values) and soil matric potential (tensiometer values) as measured in summer 1992 under the birch-pine forest BP3. Each line represents a wetting or drying event with a large number of measurements at 15 min intervals. The dotted line gives the corresponding θ/Ψ_m relationship from laboratory measurements.

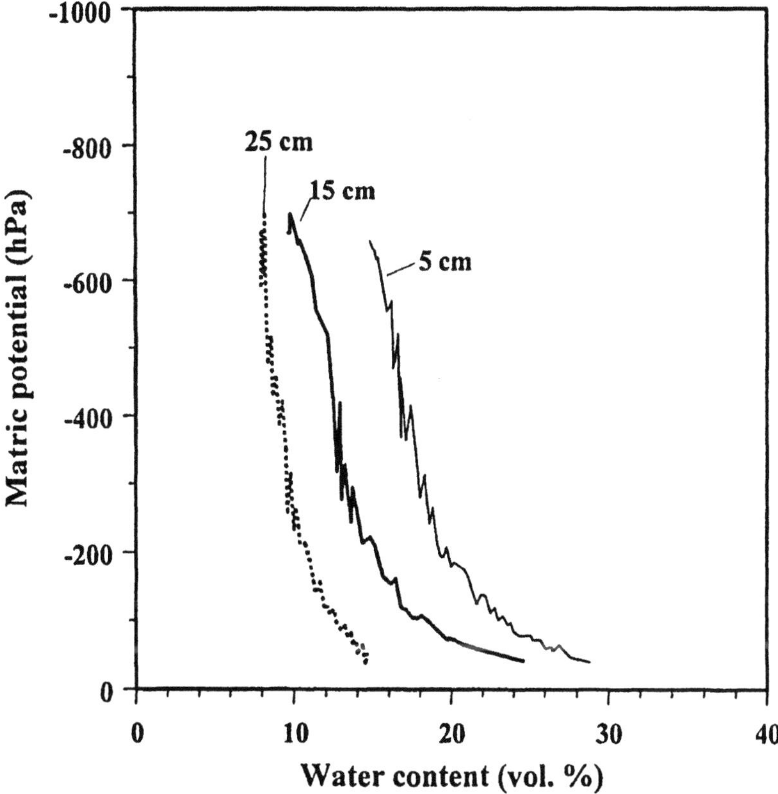

Fig. 1-8. Relationship between volumetric water content at 5, 15 and 25 cm depth (tdr values) and soil matric potential (tensiometer values) as measured in summer 1997 under the oak-beech forest OB5 (data of F. Schipka).

Laboratory measurements of unsaturated soil hydraulic conductivity show a decrease of conductivity by two or three orders for a matric potential drop between -10 hPa (close to maximum water content) and -100 hPa (less than 10 vol. % water remaining, Fig. 1-9). Additional data on the soil hydraulic conductivity at water saturation were obtained from irrigation experiments in the field (site BP 3). The downward movement of the wetting front was monitored with tdr probes and gave saturated conductivities in the range of 15 to 30 m d^{-1}, which is more than 10times higher than laboratory conductivity values obtained at -10 hPa.

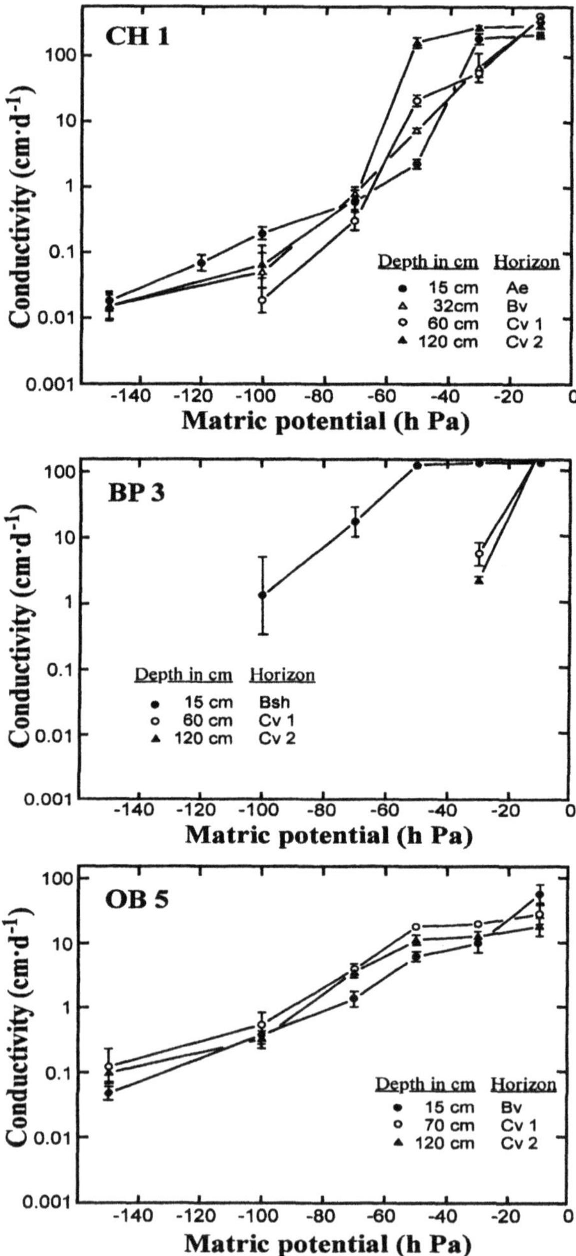

Fig. 1-9. Relationship between soil matric potential and unsaturated hydraulic conductivity of soil samples at three or four depths under heathland (site CH1), birch-pine forest (BP3) and oak-beech forest (OB5) as determined with hanging water columns in the laboratory (mean of 5 replicates each).

1.4.1.2 Organic layers

Humus accumulation considerably influences the availability and turnover of soil water in the course of the heathland-to-forest succession (Leuschner 1998, 2001, Leuschner and Rode 1999). The amount of water stored in the organic layers at the forest floor depends on (1) the thickness of the organic horizons, 2) the water retention characteristics of the organic material, and (3) the hydraulic conductivity of the humus. One of the most striking changes taking place during this succession is the rapid increase in depth of the organic layers atop of the soil. The thickness of the L, Of and Oh horizons was found to increase from 25 to 75 mm from the heathland to the birch-pine forest stage, and grew further to 90 mm in the late-successional oak-beech forest stage (Table 1-4).

Table 1-4. Thickness (in mm) of the L-, Of- and Oh-layers and density of dry mass per volume (in g cm^{-3}) of the organic profiles under heathland, birch-pine forest and oak-beech forest.

Community	T^a	n^b	L	O_f	O_h	Profile^c	Density of O_f	Density of O_h
Calluna heathland CH1	>10	6	7	10	7	24	0.294	0.186
Calluna heathland CH2	>10	6	10	16	10	36	0.181	0.304
Birch-pine forest BP4	c.45	6	18	44	7	69	0.154	0.129
Birch-pine forest BP3	c.85	12	10	63	9	86	0.105	0.408
Oak-beech forest OB5^d	>300	24	20	58	14	92	0.154	0.464
Oak-beech forest OB6^d	>300	12	18	67	32	117	0.119	0.446

^{a}Approximate time span of humus accumulation since last major disturbance (sodcutting, burning) (in years), ^{b}number of profiles investigated, ^{c}L- and O_{fh}-layers, ^{d}Putative terminal stages of succession.

Not only the horizon depth but also the structure of the organic material changes during succession, which influences the water retention capacity. Very low volume densities of the fresh L material (0.02-0.12 g cm^{-3}) contrast with much higher values for the humificated O_h material (0.13-0.46 g cm^{-3}, Table 1-4). Thick O_h-layers are confined to the late-successional oak-beech forest community, a fact that significantly contributes to the very high water storage capacity of humus derived from *Fagus* and *Quercus* litter as compared to *Betula, Pinus* or *Calluna* litter.

The water content/matric potential relationship (θ/Ψ_m curve) as determined by desorption in the laboratory gave a maximum water content (θ_{max}) value of about 90 vol. % for organic material of the O_f- and O_h-layers of the two forest communities. This is twice as high as the θ_{max} value of the quartzitic, medium-grained sand that is situated below the forest floor (Fig. 1-10). Moreover, the organic material retained two to four times more water in the plant-available matric potential range (-2.5 to -15000 hPa) than the sand. These properties favour root water uptake especially in the lower, more decomposed layers of the organic profile (O_h-horizon) and stimulate fine root growth.

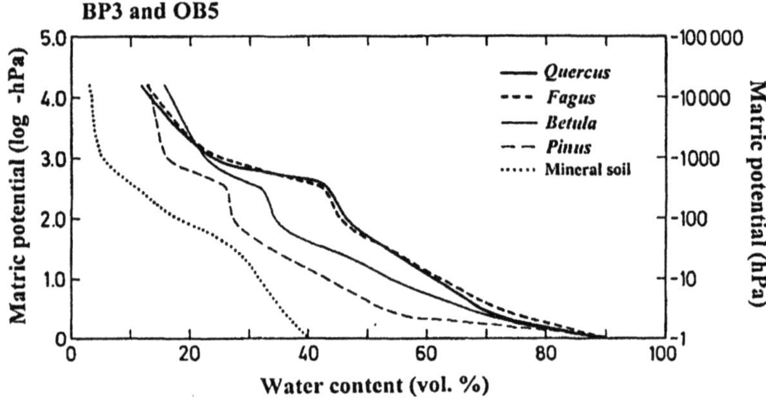

Fig. 1-10. Relationship between volumetric water content and soil matric potential of organic layer samples (Ofh horizon, laboratory desorption) taken under *Quercus* or *Fagus* trees (oak-beech forest), and *Betula* or *Pinus* trees (birch-pine forest, five replicates each). For comparison, the characteristic of the mineral topsoil is given (dotted line).

The θ/Ψ_m curve of humus material differs markedly among the four litter types (tree species) investigated: humus derived from either *Fagus* or *Quercus* debris (late-successional community) showed very similar desorption characteristics, but *Pinus* humus gave retention curves that were markedly shifted to lower water contents in the physiologically important potential range (Fig. 1-10). The amount of plant-available water, therefore, was by 20 vol. % smaller for *Pinus* humus (42.5%) than for *Quercus* or *Fagus* humus (about 60%, Table 1-5). *Betula* humus showed θ/Ψ_m curves that were intermediate between those of *Pinus* and *Fagus*. Humus of all four species retained a large quantity of water in the non-root-extractable range (water < -15000 hPa) with no clear differences between *Fagus, Quercus, Betula* and *Pinus*. Although the xeric *Calluna* humus was not investigated, it can be assumed that its hydrologic properties are similar to those of pine humus.

Infiltration experiments with undisturbed forest floor sods gave empirical relationships between the amount of rainfall and the resulting seepage loss to the mineral soil. These relationships are influenced by (1) the wetting characteristics of the humus material, i.e. the tendency of the matrix to absorb a portion of the infiltrating water, and (2) the hydraulic conductivity of the organic material. Both properties are strongly dependent on the initial water content of the humus material. Leuschner (1998, and unpubl. results) used quadratic equations to describe the water absorption of organic material and the resulting drainage as a function of throughfall rate and initial water content (wetting characteristics) for different humus types. They allow the calculation of the saturated water content θ_s (i.e. the water content immediately after a saturating infiltration) and the water retention capacity θ_r (i.e. the difference between saturated water content θ_s and initial water content) for various water contents of humus material derived from *Fagus* or *Pinus* litter.

Table 1-5. Average depth and hydrologic characteristics of the organic profiles under *Betula* and *Pinus* trees (birch-pine forest BP3), and *Quercus* and *Fagus* trees (oak-beech forest OB5) expressed as volumetric water content (in %) or water reserves (in mm per profile) (derived from gravimetric sampling; θ_{max}-porosity, i.e. all space filled with water; 'non-root-extractable' - water held at matric potentials <-15000 hPa; 'plant-available' - water held between -2.5 and -15000 hPa).

		Betula (Site BP3)	*Pinus* (Site BP3)	*Quercus* (Site OB5)	*Fagus* (Site OB5)
Profile depth	mm	87	85	104	80
Maximum water content θ_{max}	vol. %	88.6	91.6	90.5	90.0
	mm	77.1	77.9	94.1	72.0
Non-root-extractable water	vol. %	15.8	13.2	12.1	13.3
	mm	13.8	11.2	12.6	10.6
Plant-available water	vol. %	53.9	42.5	59.9	62.2
	mm	46.9	36.1	62.3	49.8

For the organic material under *Fagus*, θ_s is by 30% smaller for initially dry humus (10 mm water content) than for wet humus (31.5 mm). On the other hand, dry material has a five times higher water retention capacity and, as a result, releases less seepage water to the mineral soil than wetter material. The saturating rainfall (throughfall) amount that is needed to reach θ_s is much higher, however, for dry humus than for initially wet humus. Thus, large seasonal fluctuations of the humus water content result in considerable temporal variations in both θ_s and θ_r and, consequently, in the amount of water that percolates to the mineral soil under a given infiltration rate. More details on the hydrology of the organic layers of the successional communities (in particular of the oak-beech forest site) are given in Leuschner (1998).

1.4.2 Soil water content

1.4.2.1 Mineral soil

Two-and-a-half years of gravimetric soil moisture recording revealed an astonishing vertical gradient in volumetric water content in the profiles of all three communities: it was negative, i.e. the water content decreased with soil depth in summer. The mineral topsoil at 0-5 cm depth had a significantly higher water content than lower horizons at 10-15 or 55-60 cm depth in all seasons except for extremely dry summer periods (Fig. 1-11). While the topsoil water content exceeded 30 or 40 vol. % during winter periods, the moisture content of the subsoil typically ranged between 5 and 15%, and only exceptionally was higher than 20% in winter. Thus, field capacity (conventionally: water held at -100 hPa, Ehlers 1996) or maximum water content (θ_{max}) never were reached in the subsoil of the three successional communities. The large and significant downward gradient in soil moisture is also visible from annual (or summer) means of soil water content (Table 1-6). The vertical differences were larger for profiles under oak-beech forest than for profiles under heathland or birch-pine forest. Detailed studies with both tdr probes and gravimetric sampling in a soil pit of the oak-beech forest in October 1991 confirmed the vertical decrease of θ. The data in Figure 1-12 reveal a soil moisture minimum at 40 to 70 cm depth and a subsequent increase deeper in the profile. The decrease in water content with depth in the upper 40 cm of the profile is unexpected for two reasons: (i) moist winter periods obviously saturated the topsoil and led to high infiltration rates (as is visible from modelled water fluxes in the soil) but did not lead to field capacity in the subsoil, and (ii) the higher water content of the topsoil persisted even during many summer periods despite a much higher fine root density near the surface and, thus, elevated root water uptake rates compared to the subsoil.

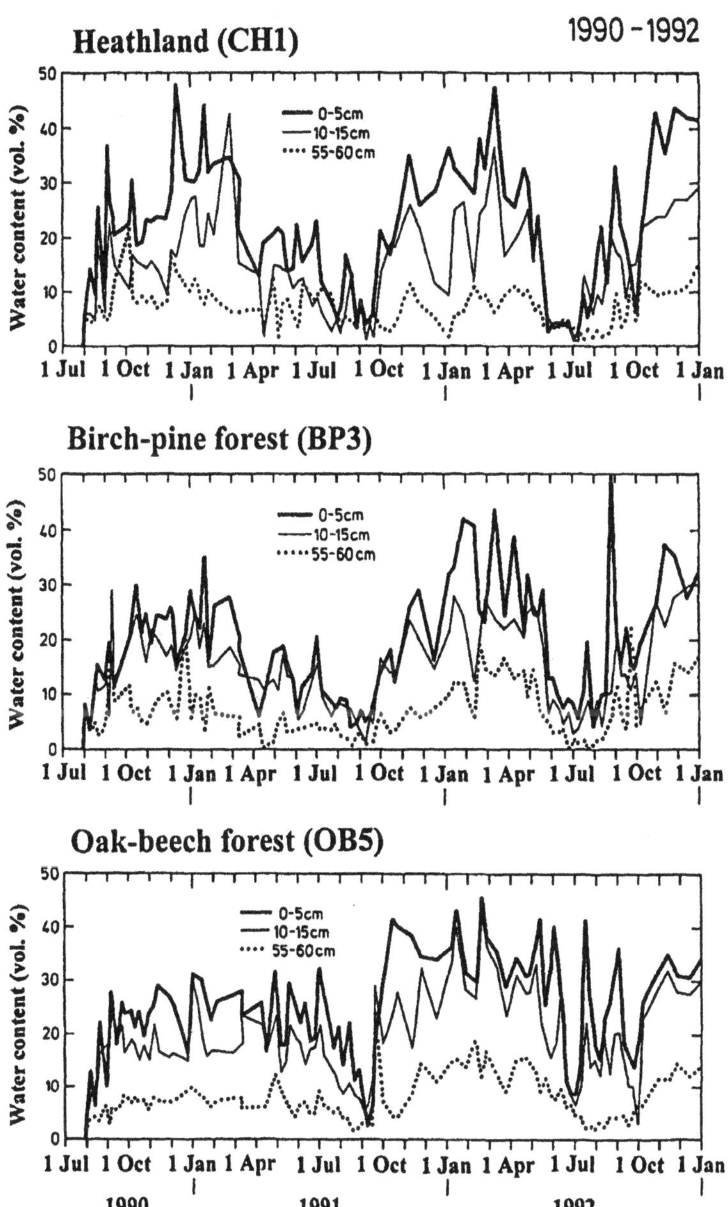

Fig. 1-11. Seasonal course of volumetric water contents at three soil depths in profiles under heathland (site CH1), birch-pine forest (BP3) and oak-beech forest (OB5) during 1990 to 1992 (3-5 replicate samples analysed by gravimetry).

Table 1-6. Mean volumetric water content (in %) of three mineral soil horizons in profiles under heathland (site CH1), birch-pine forest (BP3) and oak-beech forest (OB5) in the periods 1 January-31 December, 1991 or 1992, and 15 May-20 September, 1991 or 1992 (n = 3-5 gravimetric samples, coefficients of variation (v = STD/mean) in parentheses). Roman letters indicate significant differences (p < 0.05) between different communities for the 0-5 cm depth in the full year or summer periods, greek letters mark differences between the communities at 10-15 cm depth, and numbers indicate differences at 55-60 cm depth. Differences between two depths for a given community were all significant except for the 0-5 vs. 10-15 cm difference at the heathland in summer 1992.

Site	Depth	1991		1992	
	(cm)	1 Jan-31 Dec	15 May-20 Sep	1 Jan-31 Dec	15 May-20 Sep
CH1	0-5	19.4 $(0.09)^a$	12.8 $(0.13)^a$	22.4 $(0.11)^a$	11.9 $(0.19)^a$
CH1	10-15	13.4 $(0.11)^\alpha$	8.1 $(0.13)^\alpha$	15.9 $(0.09)^\alpha$	10.1 $(0.16)^\alpha$
CH1	55-60	6.8 $(0.07)^1$	6.5 $(0.12)^1$	8.3 $(0.12)^1$	3.9 $(0.13)^{1,2}$
BP3	0-5	15.2 $(0.09)^a$	10.3 $(0.12)^a$	22.4 $(0.11)^a$	15.1 $(0.17)^a$
BP3	10-15	12.5 $(0.08)^\alpha$	8.7 $(0.13)^\alpha$	16.0 $(0.09)^\alpha$	9.6 $(0.15)^\alpha$
BP3	55-60	4.3 $(0.10)^2$	3.3 $(0.11)^2$	6.3 $(0.09)^{1,3}$	3.7 $(0.20)^1$
OB5	0-5	23.4 $(0.07)^b$	18.9 $(0.11)^b$	28.1 $(0.06)^b$	24.7 $(0.10)^b$
OB5	10-15	17.5 $(0.06)^\beta$	13.5 $(0.09)^\beta$	22.5 $(0.07)^\beta$	16.6 $(0.08)^\beta$
OB5	55-60	6.9 $(0.09)^1$	5.0 $(0.10)^1$	8.8 $(0.09)^{2,3}$	5.3 $(0.13)^2$

The seasonal dynamics of soil moisture in the three communities are characterised by a moisture recharge period in September and October, winter maxima of water content in January and February, and a decrease of moisture starting as early as in March or April (Fig. 1-11). Soil moisture minima may reach θ values < 2 vol. % in the sandy material during extended drought periods in summer. The recharge of the soil moisture reserves was less complete in winter 1990/1991 than in 1991/1992 (Fig. 1-13).

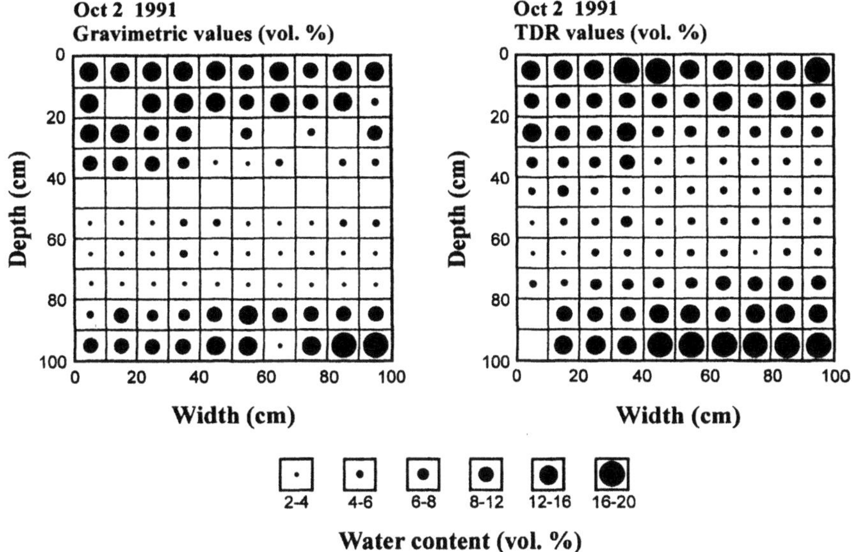

Fig. 1-12. Volumetric water content of the soil in a profile wall of 1 m depth and 1 m width under the oak-beech forest on October 2, 1991, as determined gravimetrically (left) and by tdr probes (right).

Larger differences in soil moisture content existed between topsoil and subsoil within a given profile than among the different profiles under the three communities. In the lower profile at 55-60 cm depth, only weak seasonal fluctuations of the water content occurred in the 30-months study period (Fig. 1-13: lower panel). In contrast, large seasonal fluctuations occurred in the topsoil at 0-5 cm depth (Fig. 1-13: upper panel). In this horizon, significant differences existed between the mean water content of the heathland and the oak-beech forest, and that of the birch-pine forest and the oak-beech forest (Table 1-6). Thus, the topsoil of the oak-beech forest (0-5 and 10-15 cm) had a higher moisture content than both the heathland and the birch-pine forest profiles during most summer and also winter periods in the three years. In contrast, the topsoils under heathland and birch-pine forest differed not significantly in their water contents. In the subsoil (55-60 cm), the birch-pine forest showed significantly smaller values compared to the other communities in 1991 but not in 1992.

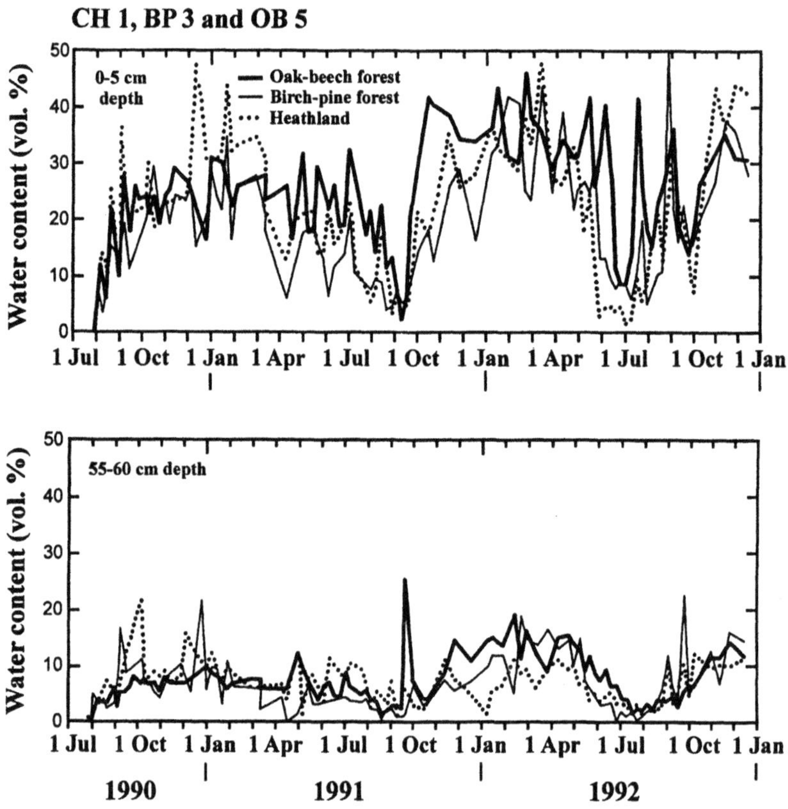

Fig. 1-13. Seasonal course of volumetric water content at 0-5 cm depth (upper panel) and 55-60 cm depth (lower panel) in profiles under heathland (site CH1), birch-pine forest (BP3) and oak-beech forest (OB5) during 1990 to 1992 (3-5 replicate samples analysed gravimetrically).

1.4.2.2 Organic layers

The 8 to 10 cm thick mor profiles of the oak-beech forest contained considerable water reserves not only during wet seasons but also during periods of summer drought. While winter values peaked at 50 vol. % under *Quercus* trees, summer values ranged between 25 and 40 vol. % in wet periods and reached minima of 18% in periods of drought (Fig. 1-14). With minima of 10 vol. % the organic horizons under *Fagus* trees were somewhat drier than those under neighbouring *Quercus* trees in the OB5 stand. Pine humus, which consists mainly of the hydrophobic *Pinus* needles, reached summer minima < 5 vol. % (Fig. 1-14).

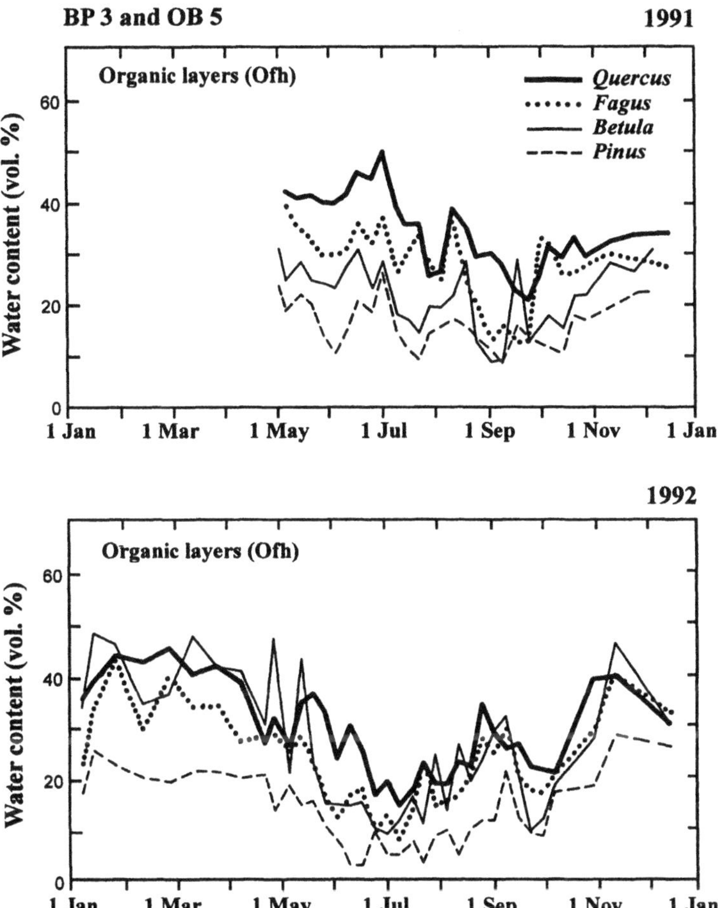

Fig. 1-14. Seasonal course of the volumetric water content of the organic layers (O_{lfh}-horizons) under *Quercus* or *Fagus* trees (oak-beech forest), or *Betula* or *Pinus* trees (birch-pine forest) during 1991-1992 (means of 4 to 8 replicate samples analysed by gravimetry).

Betula humus showed moisture contents during summer that were more or less intermediate between *Fagus* and *Pinus* (Table 1-7), but this material can reach high water contents during winter. The differences among the tree species are larger when the average water reserves in the organic layers are considered. During summer they were more than three times larger under *Quercus* in the late-successional forest than under *Pinus* in the pioneer forest due to both differences in wetting characteristics and profile thickness (Tab. 1-7). Maximum storage peaked at 45 mm under *Quercus* in winter but reached only 22 mm under *Pinus* (Fig. 1-15). No data are available for the only 24 mm thick organic profile in the heathland.

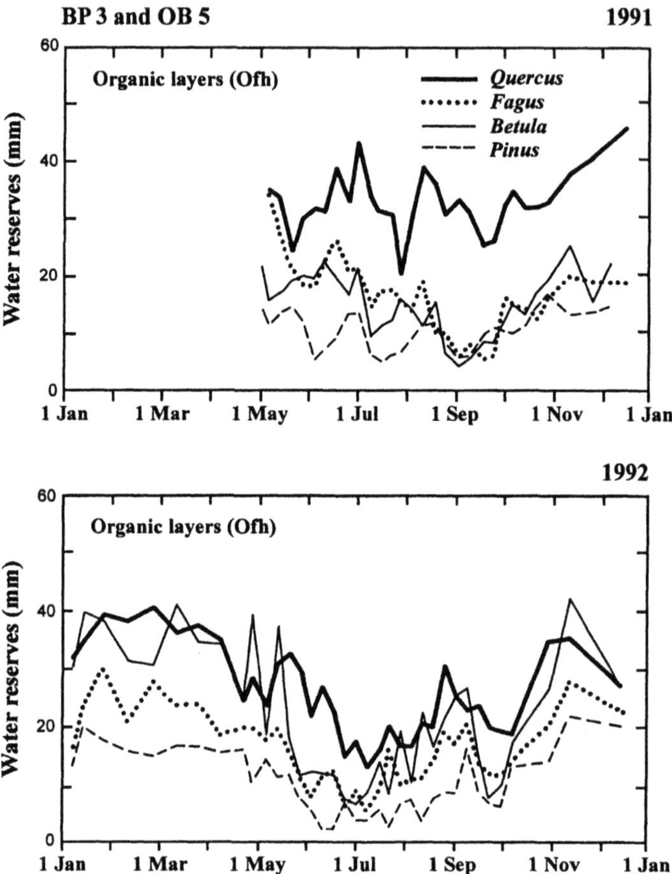

Fig. 1-15. Seasonal course of the water reserves stored in the organic layers (O_{lfh}-horizons) under *Quercus* or *Fagus* trees (oak-beech forest), or *Betula* or *Pinus* trees (birch-pine forest) during 1991-1992 (means of 4 to 8 replicate samples analysed by gravimetry).

Table 1-7. Average volumetric water content (in %) and water reserves (in mm) in organic profiles under *Betula* or *Pinus* trees (site BP3), and *Quercus* or *Fagus* trees (site OB5) in summer periods of 1991 and 1992.

	14 May-17 Sep, 1991		18 May-20 Sep, 1992	
	vol. %	mm	vol. %	mm
Betula	21.2	14.6	18.3	15.2
Pinus	15.5	9.3	9.0	7.0
Quercus	36.8	32.2	24.3	21.8
Fagus	28.1	16.4	17.8	12.5

1.4.2.3 Total soil profile

Conventionally, water held in the range between -1.5 MPa and -300 or -100 (or -60) hPa is termed 'plant-available water' (PW, Ehlers 1996). On the other hand, water bound at matric potentials >-100 hPa is thought to leave the rooted soil volume rapidly by percolation in the macropores and thus is unavailable for plant uptake. However, tensiometer measurements during the winter months showed that water held at potentials between -100 and -20 hPa apparently remained for periods of several days to weeks in the rooted horizons at the three study sites. It did not leave the profile as drainage water as is predicted by theory (Fig. 1-16). Therefore, a fraction of the water held at matric potentials > -100 hPa also must be considered as plant-available in the profiles studied. Medium-grained sandy soils may thus have larger soil water reserves than previously thought. As a consequence, all water bound in the range of -2.5 hPa to -1.5 MPa was considered as plant-available in this study.

Forty to 50 mm of plant-available water were stored at average in 1991 in the mineral soil profiles (0 - 70 cm depth) under the three successional communities (Table 1-8). During mid-summer 1991 (15 May to 20 September), this pool was reduced to 23-33 mm. Winter maxima of PW peaked at 100 to 150 mm, summer minima periodically approached zero. In fact, no plant-available water at all was recorded in the mineral soil profiles for a short dry period early in September, 1991 (Fig. 1-17). These data on soil water depletion demonstrate the physiological importance of periodic water limitation in all three communities of the heathland-to-forest succession.

Profiles under the oak-beech forest contained significantly larger water reserves in summer and winter 1991 (and also in 1992, data not shown) than the profiles of the two other vegetation types (Tab. 1-8). This difference existed despite the fact that the geological substrates under the three successional communities are similar. Possible explanations are (i) the lower density of the topsoil in the oak-beech forest (see Ch. 1.4.1.1), and (ii) the evergreen habit of the plants in heathland and pioneer forest with comparably high transpiration and interception losses during winter.

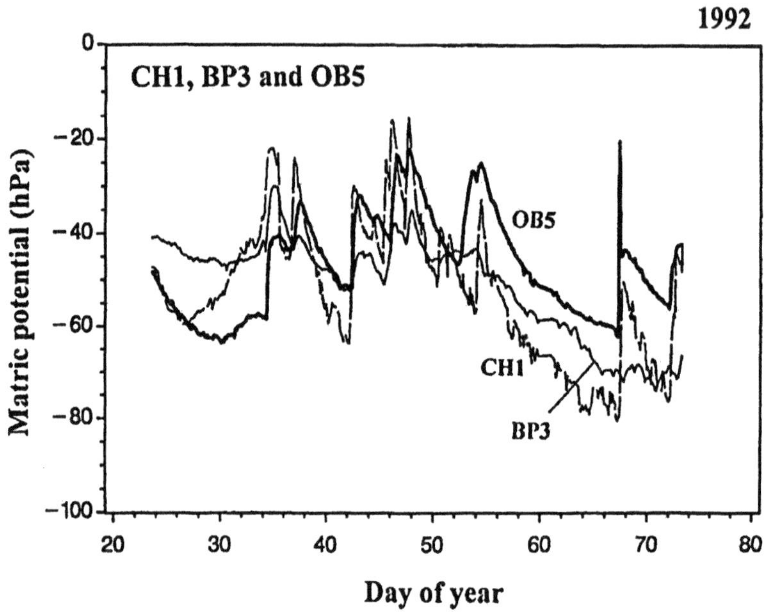

Fig. 1-16. Seasonal variability of the matric potential in the mineral soil under the heathland (dashed line), birch-pine forest (thin line) and oak-beech forest (bold line) from January to March 1992 (means of 4 to 5 tensiometer readings per site at 15 and 60 cm depth).

In the oak-beech forest community, water stored in the organic layers (mid-summer average: 12.1 mm) contributed significantly to the soil water pools. In this community, the organic layers play an important role in forest hydrology not only in wet winter months but also in dry summer periods (Leuschner 1998). For example, the thick organic profile stored several mm of plant-available water even in dry August/September 1991 when the sandy mineral soil profile had completely lost its plant-available water (Fig. 1-17: bottom panel).

Eight to 21% of the plant-available water reserves held in the profiles to 70 cm depth referred to water in the organic layers in the oak-beech forest. The remaining water was stored in the mineral soil. These data emphasize the hydrological importance of the organic horizons in the late-successional community. In the birch-pine forest, in contrast, water availability was improved only slightly due to the existence of a forest floor water pool (Tab. 1-8). The average pool size in summer was only 1.6 mm, which equals 4 to 7% of the profile total.

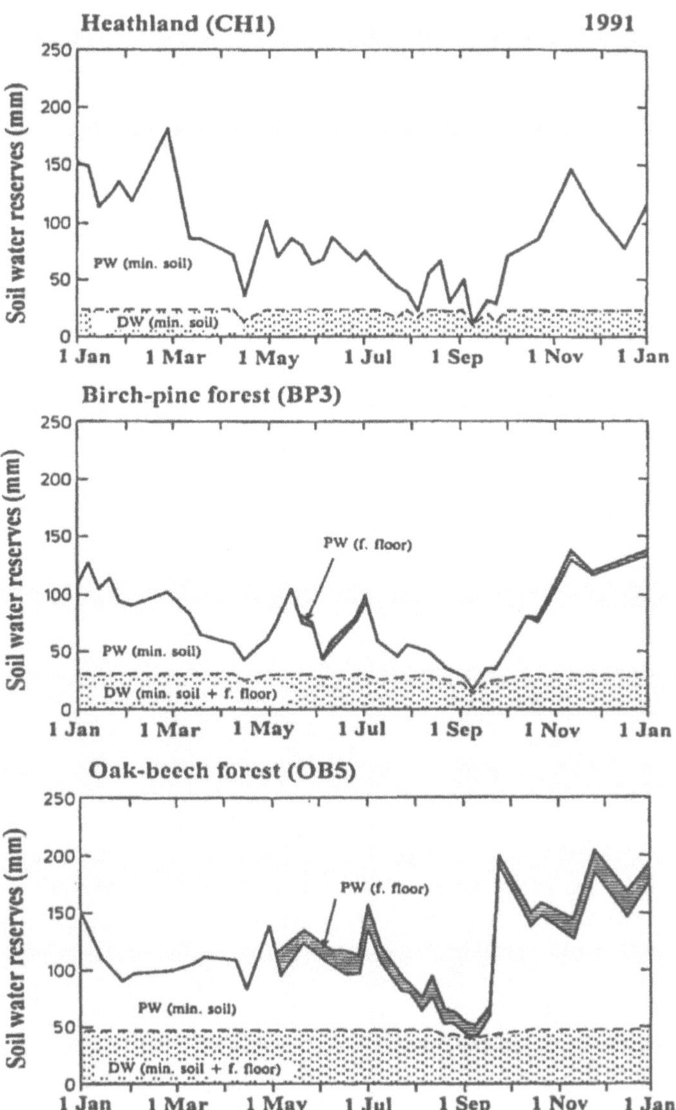

Fig. 1-17. Cumulative water reserves in soil profiles under heathland (CH1), birch-pine forest (BP3) and oak-beech forest (OB5) in 1991 (soil depth 0-70 cm). Given is 'plant-available' water (PW, water held in the potential range between -2.5 hPa and -1.5 MPa) and 'non-root-extractable' water (DW, < -1.5 MPa). No measurements were conducted in January-April in the forest floor of sites BP3 and OB5.

Table 1-8. Average water reserves (in mm) in the mineral soil profile (0-70 cm depth), in the organic horizons (L-, O_f-, O_h-layers) and in the total soil profile (i.e. mineral plus organic horizons) during mid-summer 1991 (15 May-20 Sep), and during the year 1991 (1 Jan-31 Dec) under the three successional communities. Given are total soil water content (TW) and plant-available water content (PW, water held between -2.5 hPa and -1.5 MPa); n.d. - not determined. Average coefficients of variation (STD/mean) were 0.18, 0.17 and 0.14 for the water content data of the mineral soils of heathland, birch-pine forest and oak-beech forest, respectively, and 0.23 and 0.15 for the organic layers of birch-pine forest and oak-beech forest, respectively; n = 4 to 8 samples.

	Mineral soil profile		Organic horizons		Total soil profile	
	TW	PW	TW	PW	TW	PW
			15 May-20 Sep			
Calluna heathland	56.6	27.7	c. 4	c. 2	60.6	29.7
Birch-pine forest	42.5	23.3	11.9	1.6	54.4	24.9
Oak-beech forest	73.2	33.0	24.3	12.1	97.5	45.1
			1 Jan -1 Dec			
Calluna heathland	78.4	46.8	n.d.	n.d.	n.d.	n.d.
Birch-pine forest	57.7	39.7	n.d.	n.d.	n.d.	n.d.
Oak-beech forest	96.2	50.7	n.d.	n.d.	n.d.	n.d.

In the pioneer forest, the organic layers are thinner (7 to 8 cm vs. 8 to 10 cm) and more hydrophobic compared to the oak-beech forest community (Leuschner 1998). No significant water storage was observed during summer in the thin organic profile of the heathland (Fig. 1-17: upper panel). To summarise, the forest floor water reserves increased about sevenfold during heathland-to-forest succession and significantly contributed to plant water supply in late stages of succession.

1.4.3 Soil matric potential

1.4.3.1 Mineral soil

Continuous measurements with stem-centred arrays of tensiometers at 5 to 6 depths revealed similar seasonal patterns of the soil matric potential for the pioneer and the late-successional forest communities. Matric potential courses under the *Calluna* dwarf shrubs in the heathland were not principally different in their seasonality from the patterns found under the two forest communities. Periods of high rainfall and low evapotranspiration (typically from November to April) are characterised in the entire profile by potentials in the range of -20 to -100 hPa close to water saturation. Vertical potential gradients were small and varied between -50 and +50 hPa m^{-1}. Decreases in water potential with soil depth (i.e. positive potential gradients) are a prerequisite of downward percolation and drainage. In all three communities, this situation occurred during the winter months only. Between May and October, in contrast, the direction of the matric potential gradient reversed in all three communities with highest potentials occurring in the subsoil below the rooting horizon at 130 to 170 cm depth, and lowest potentials being found in the upper profile (Figs. 1-18 to 1-23). Thus, a local water flow divide existed in the subsoil of the profiles with gravitational downward flux prevailing in the lower profile (beyond 130 or 170 cm) and upward water movement occurring along the negative potential gradient above the divide. During periods of extended drought, the negative potential gradient was as large as -300 to -400 hPa m^{-1} with a matric potential minimum being found at 40 to 60 cm depth. Thus, matric potential minima and water content minima occurred during summer at similar soil depths (at about 50 cm below the surface) in the sandy profiles. Typically, the topsoil at 0-20 cm depth had significantly higher matric potentials and also higher water contents in summer than horizons at 40-60 cm depth. Remarkably, this anomaly in the vertical soil matric potential gradient occurred not only in periods immediately after rainfall but obviously persisted in many rainless intervals (Figs. 1-18 to 1-23). Moreover, the majority of summer rainfall events in the years 1990 to 1992 did not result in a reversal of the negative soil potential gradient but, in most cases, resulted in wetter organic layers and topsoil horizons only. As a consequence of the stable water flow divide in the subsoil, virtually no drainage occurred under the two forest communities between June and September of the three summers. It appears that rainfall events >30 mm (as in early September 1990) are necessary to reverse the potential gradient and to result in measurable deep percolation.

Profiles under heathland, birch-pine forest and oak-beech forest did not differ significantly with respect to the potential minima reached in the upper soil horizons. During summer drought, tensiometer readings sometimes exceeded -800 hPa in all three communities which is at the lower limit of the measuring range of the instruments. Irregular measurements with soil psychrometers (PCT-55, Wescor, Logan, Utah, USA) gave matric potential values as low as -0.36 MPa (heathland, August 1990) and -0.53 MPa (birch-pine forest, September 1990) which is close to the supposed threshold value of tree water uptake (Waring and Running 1998). Thus, matric potential and water content data both indicate that the plant-available water can be exhausted in the sandy soil profiles during extended summer drought periods.

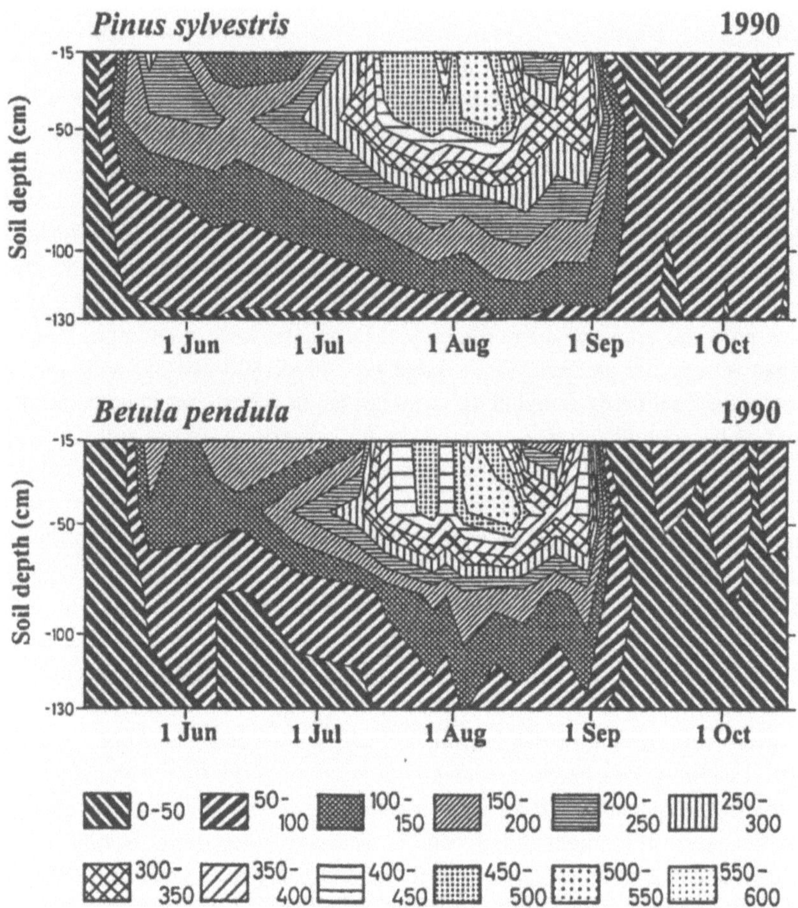

Fig. 1-18. Seasonal changes in soil matric potentials in soil profiles (15 to 130 cm depth) under *Pinus* (upper panel) or *Betula* (lower panel) in the birch-pine forest during the vegetation period 1990 (means of 7 to 22 replicate tensiometers; given are suction values = matric potentials x -1).

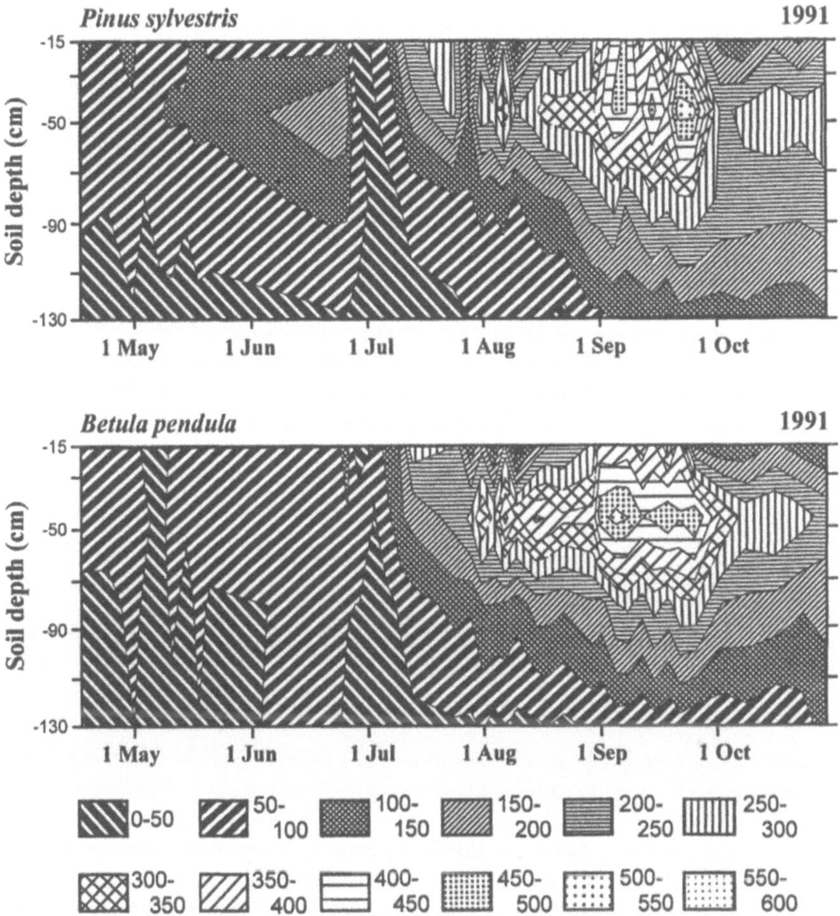

Fig. 1-19. Seasonal changes in soil matric potentials in soil profiles (15 to 130 cm depth) under *Pinus* (upper panel) or *Betula* (lower panel) in the birch-pine forest during the vegetation period 1991 (means of 7 to 22 replicate tensiometers; given are suction values = matric potentials x -1).

          C. Leuschner

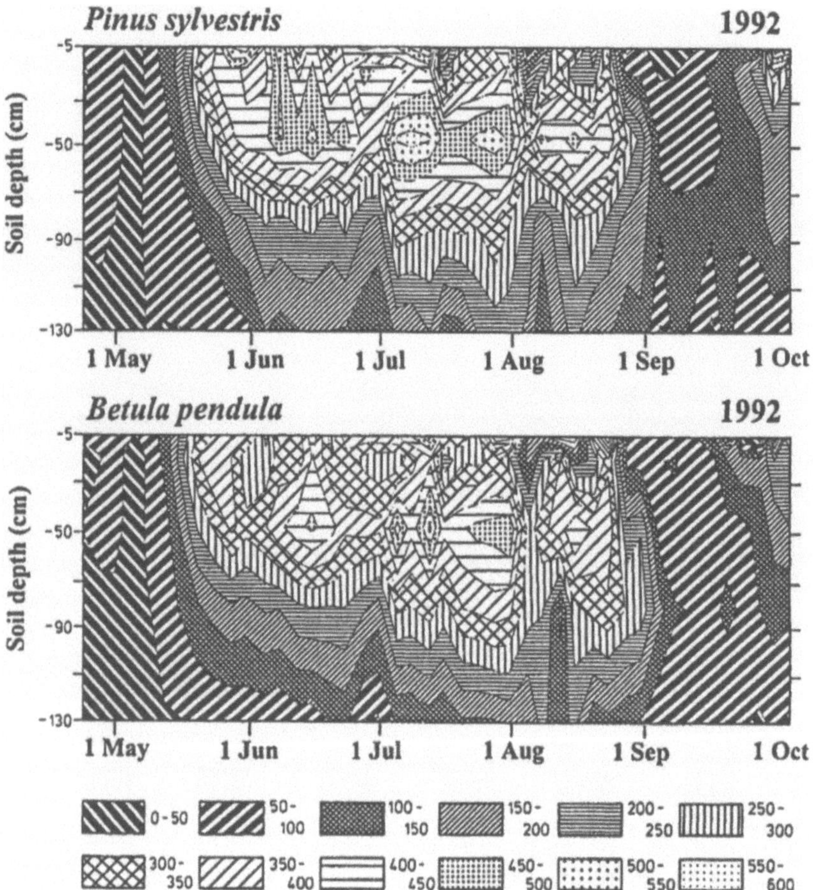

Fig. 1-20. Seasonal changes in soil matric potentials in soil profiles (5 to 130 cm depth) under *Pinus* (upper panel) or *Betula* (lower panel) in the birch-pine forest during the vegetation period 1992 (means of 7 to 22 replicate tensiometers; given are suction values = matric potentials x -1).

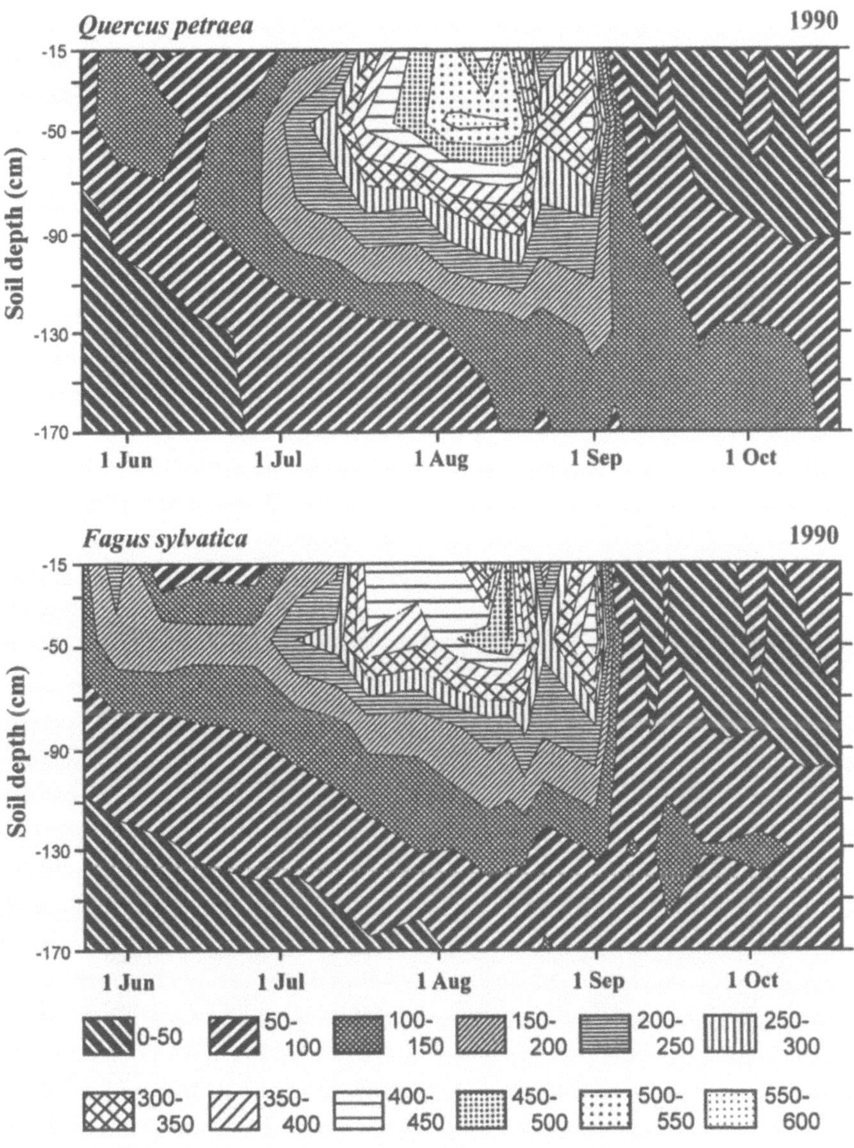

Fig. 1-21. Seasonal changes in soil matric potentials in soil profiles (15 to 170 cm depth) under *Quercus* (upper panel) or *Fagus* (lower panel) in the oak-beech forest during the vegetation period 1990 (means of 7 to 22 replicate tensiometers; given are suction values = matric potentials x -1).

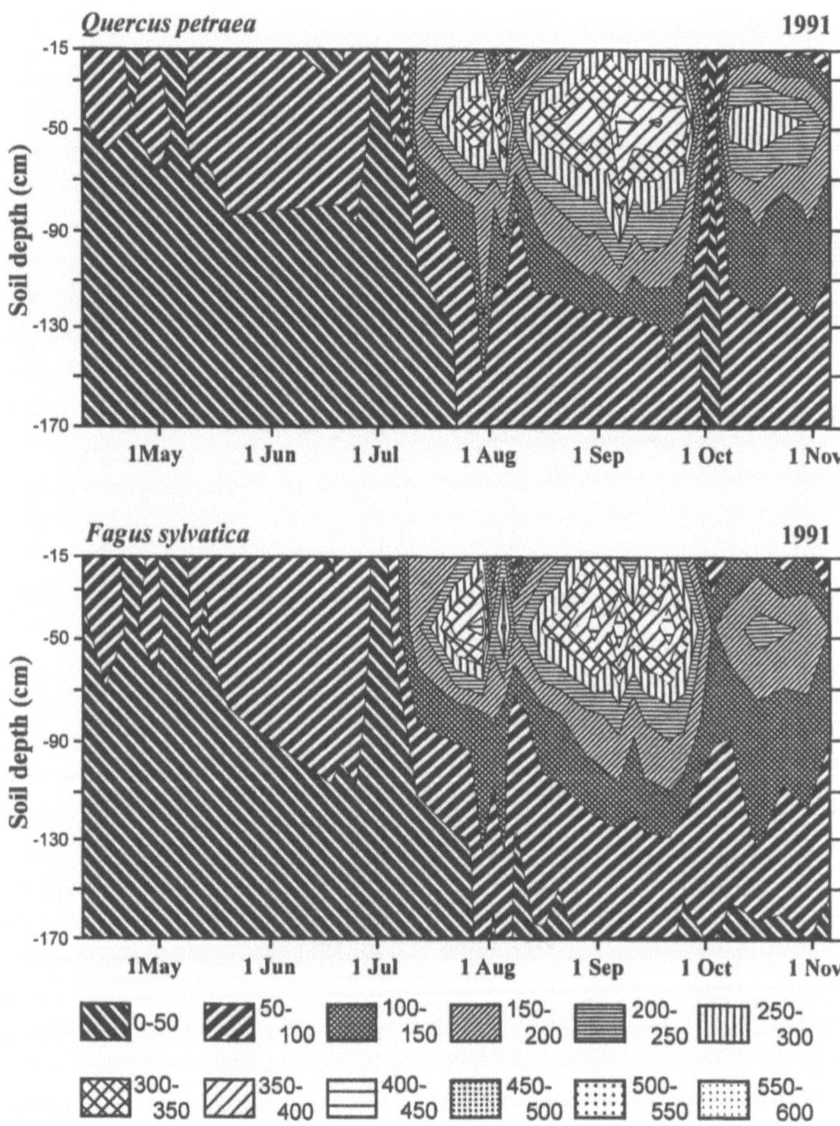

Fig. 1-22. Seasonal changes in soil matric potentials in soil profiles (15 to 170 cm depth) under *Quercus* (upper panel) or *Fagus* (lower panel) in the oak-beech forest during the vegetation period 1991 (means of 7 to 22 replicate tensiometers; given are suction values = matric potentials x -1).

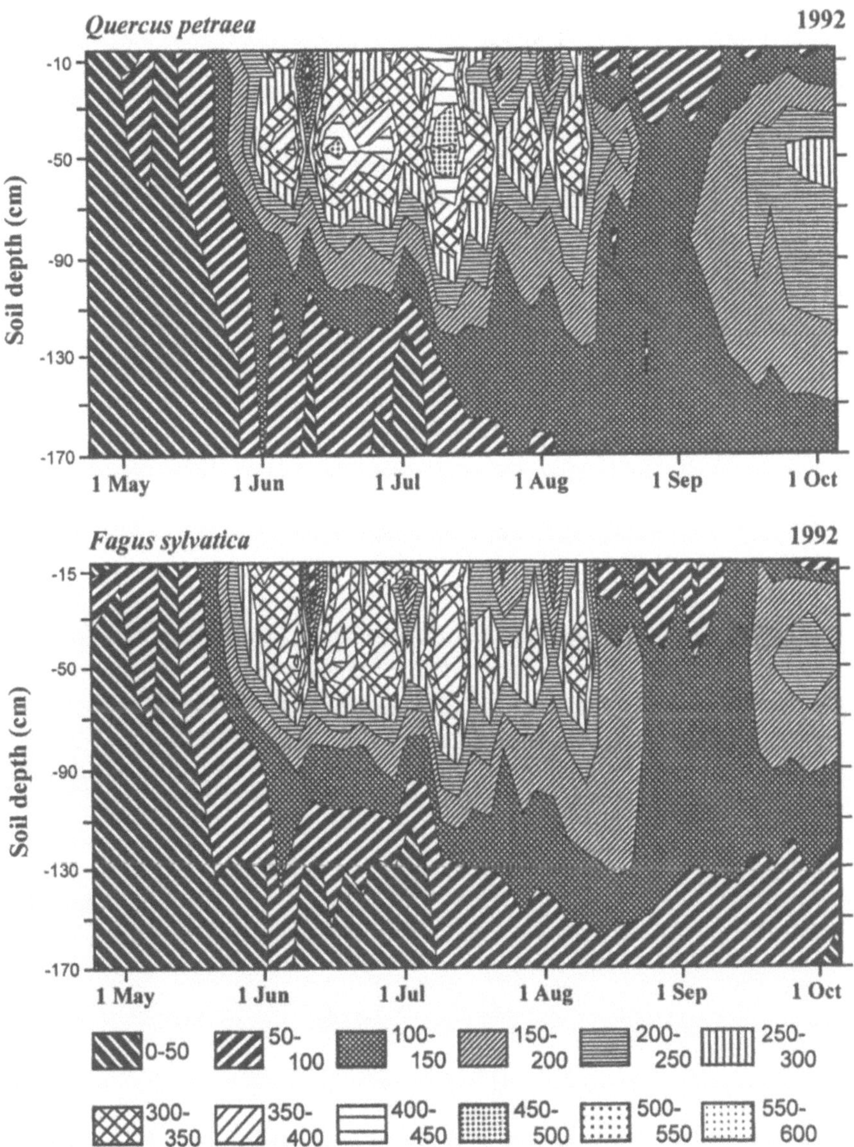

Fig. 1-23. Seasonal changes in soil matric potentials in soil profiles (5 to 170 cm depth) under *Quercus* (upper panel) or *Fagus* (lower panel) in the oak-beech forest during the vegetation period 1992 (means of 7 to 22 replicate tensiometers; given are suction values = matric potentials x -1).

Large differences existed among the three communities in the vertical extension of the water extraction horizon. Under the two forest communities, soil water depletion and the corresponding decrease in matric potentials occurred downwards to soil depths of 80 to 130 cm with minima at -400 and -200 hPa, respectively (Figs. 1-24 to 1-26). Under the heathland, in contrast, the depletion zone was restricted to the upper 60 cm of the profile (data not shown).

Between the pioneer and the late-successional forest marked differences with respect to the seasonality and vertical extension of the water depletion zone did not exist (Figs. 1-18 to 1-23). However, a detailed analysis of the seasonal matric potential courses under the coexisting tree species (*Betula* and *Pinus* in the pioneer forest, and *Quercus* and *Fagus* in the late-successional forest) indicates that at least *Fagus* and *Quercus* trees differed in their soil water depletion patterns although they are growing in direct vicinity. Soil profiles under *Quercus* stems reached lower soil matric potentials during mid-summer than profiles under *Fagus*. During several weeks in the dry period of July-October 1990 and in the moderately dry summer 1992, tensiometer readings were significantly lower in 45, 80 and 170 cm depth in the vicinity of *Quercus* stems than close to *Fagus* stems (Leuschner 1993). Differences of up to 100 hPa between the two species existed for periods of 2 to 6 weeks in the upper soil (Figs. 1-24 and 1-26). Differences were smaller in the lower profile at 170 cm depth but lasted for about three months in both summers and seemed to increase with time. During wet summer periods, however, tensiometer readings were not significantly different between the two tree species as is visible in 45 and 80 cm depth in September and October 1990 and 1992 (Figs. 1-24 and 1-26). Similarly, in the less drought-affected summer 1992, significant differences in matric potentials between *Quercus* and *Fagus* profiles were restricted to short autumn periods (Fig. 1-25).

Differences in matric potentials under the two pioneer trees *Betula* and *Pinus* occurred only during limited periods in early summer 1990 and 1991 (May/June) when profiles under *Pinus* showed significantly lower potentials than the soil under *Betula* trees (Figs. 1-24 and 1-25).

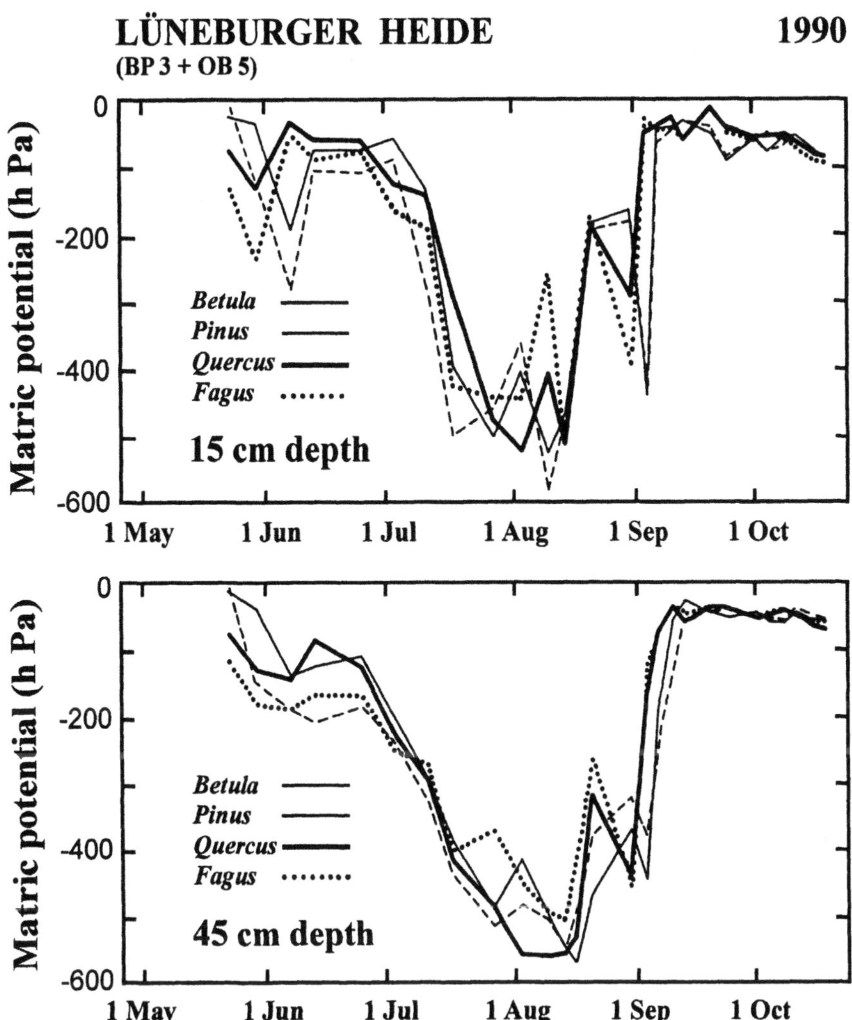

Fig. 1-24. Seasonal course of the soil matric potential at 15, 45, 80, 130 and 170 cm depth in profiles under *Betula, Pinus* (birch-pine forest site BP3), *Quercus* or *Fagus* trees (oak-beech forest site OB5) over the vegetation period 1990 (means of 7 to 22 replicate tensiometers). No data exist for 170 cm under *Pinus* and *Betula*. Significant differences between *Fagus* and *Quercus* existed in mid June and during end of July/August (45 and 80 cm), and from mid July to October (see Leuschner 1993).

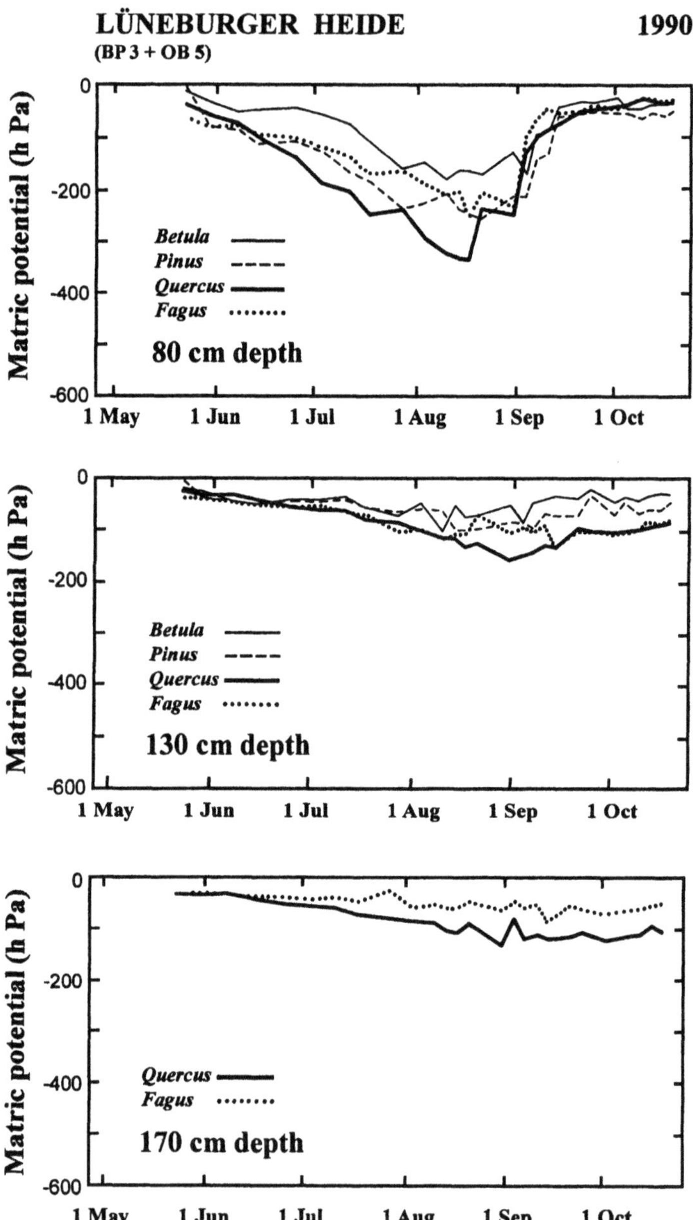

Fig. 1-24 (continued)

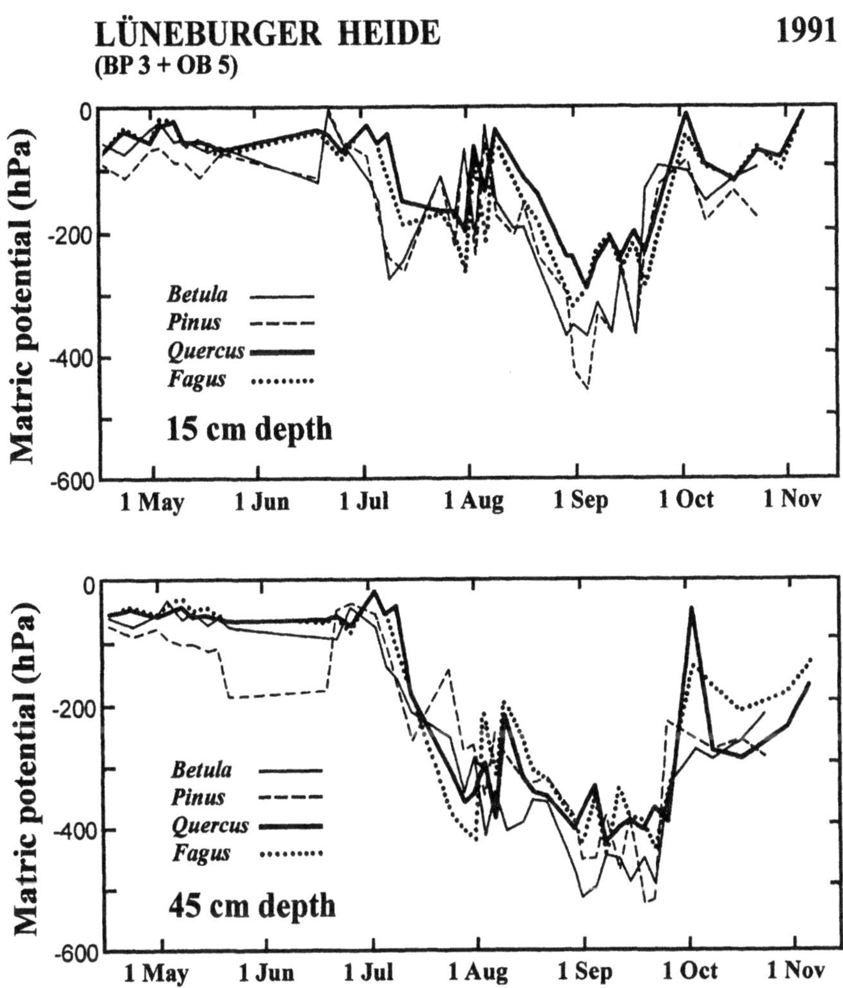

Fig. 1-25. Seasonal course of the soil matric potential at 15, 45, 80, 130 and 170 cm depth in profiles under *Betula, Pinus* (birch-pine forest site BP3), *Quercus* or *Fagus* trees (oak-beech forest site OB5) over the vegetation period 1991 (means of 7 to 22 replicate tensiometers). No data exist for 170 cm under *Pinus* and *Betula*.

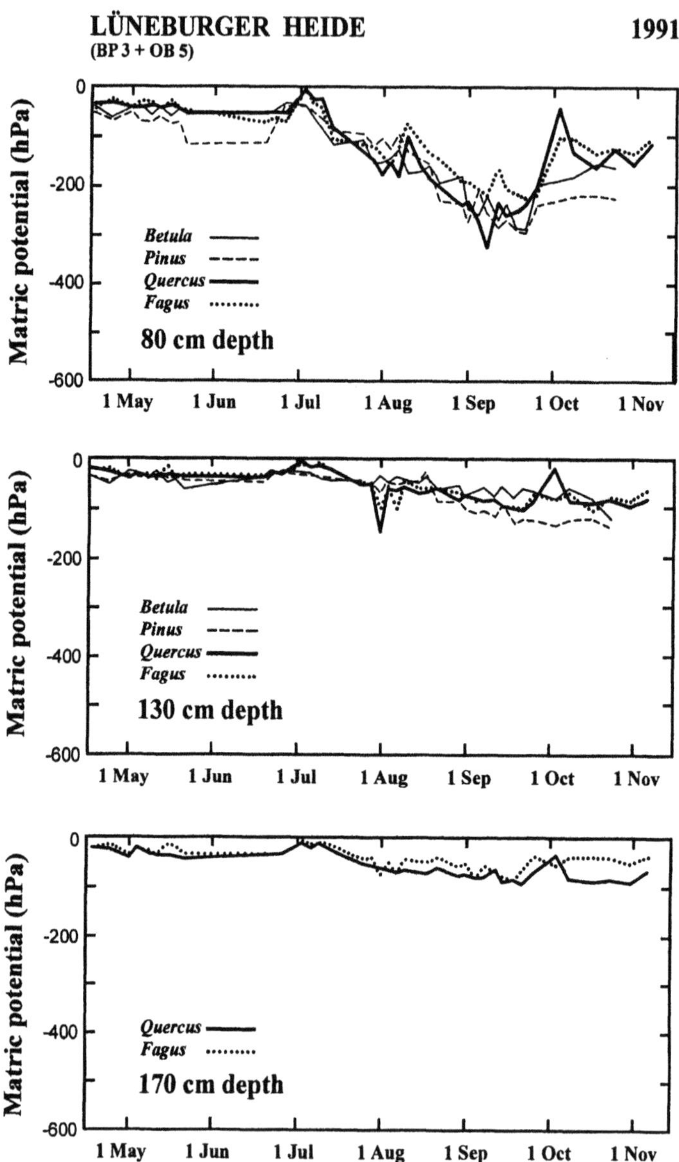

Fig. 1-25 (continued)

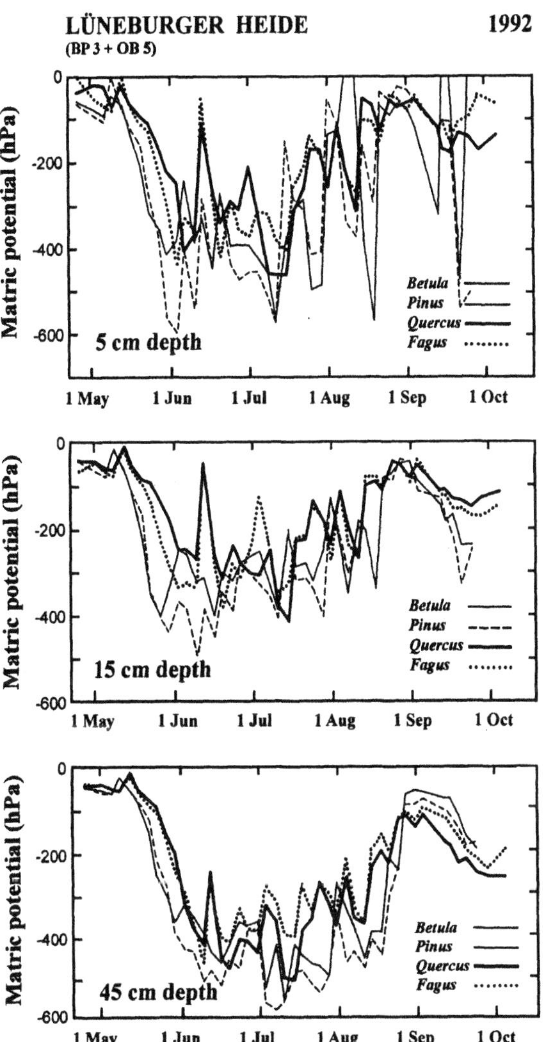

Fig. 1-26. Seasonal course of the soil matric potential at 5, 15, 45, 80, 130 and 170 cm depth in profiles under *Betula, Pinus* (birch-pine forest site BP3), *Quercus* or *Fagus* trees (oak-beech forest site OB5) over the vegetation period 1992 (means of 7 to 22 replicate tensiometers). No data exist for 170 cm under *Pinus* and *Betula*. Significant differences between *Fagus* and *Quercus* existed in mid July (45 cm), from mid September to October (130 cm), and from August to October (170 cm).

 C. Leuschner

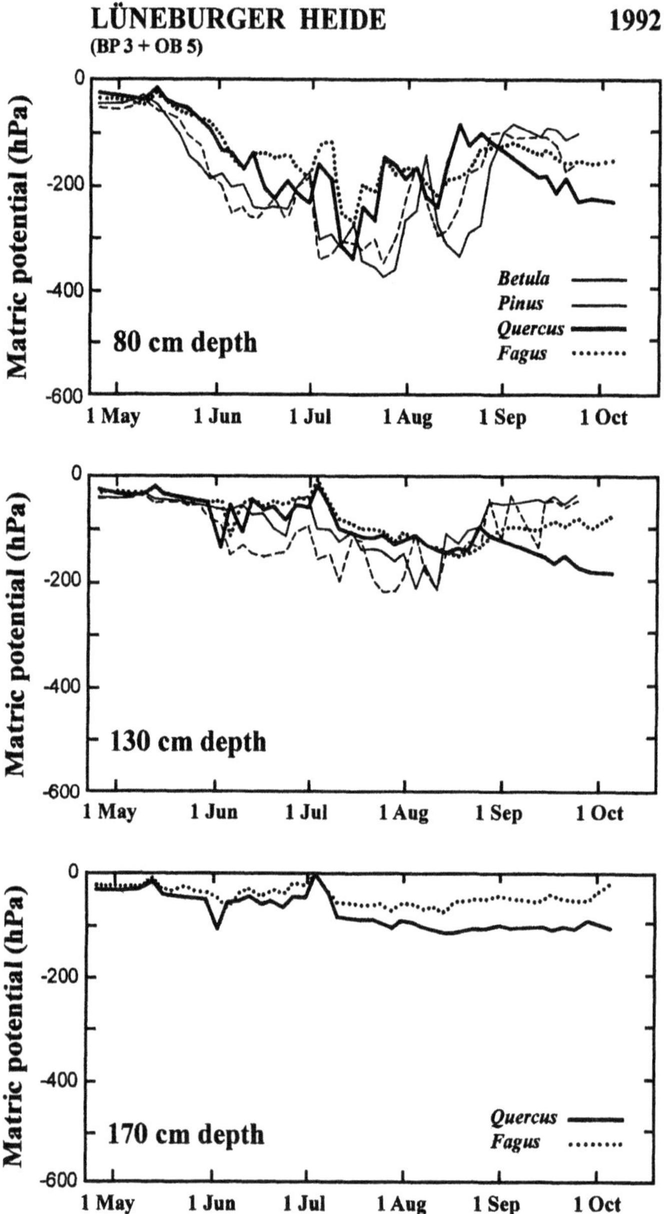

Fig. 1-26 (continued)

1.4.3.2 Organic layers

In the literature, only few attempts were made to measure the matric potential of the organic soil horizons. Due to inadequate contact to the matrix, tensiometers may be applicable only in the more compact Oh-layers (Hölzer 1982) or the uppermost mineral soil horizons. The results of soil psychrometer measurements may also be questionable. However, the forest floor typically contains high fine root densities in forest ecosystems on acid soils and, thus, may be far more important for water and nutrient uptake than most mineral soil horizons. In Figure 1-27, water potential measurements with psychrometers (Wescor PCT-55 chambers placed in the Ofh horizon) and tensiometers (positioned in the uppermost mineral soil at 2 cm depth) are depicted together with matric potentials derived from laboratory θ/Ψ_m curves (see Fig. 1-10), and water content data of the organic layers under *Betula*, *Pinus*, *Quercus* and *Fagus* trees in the pioneer or late-successional forest communities.

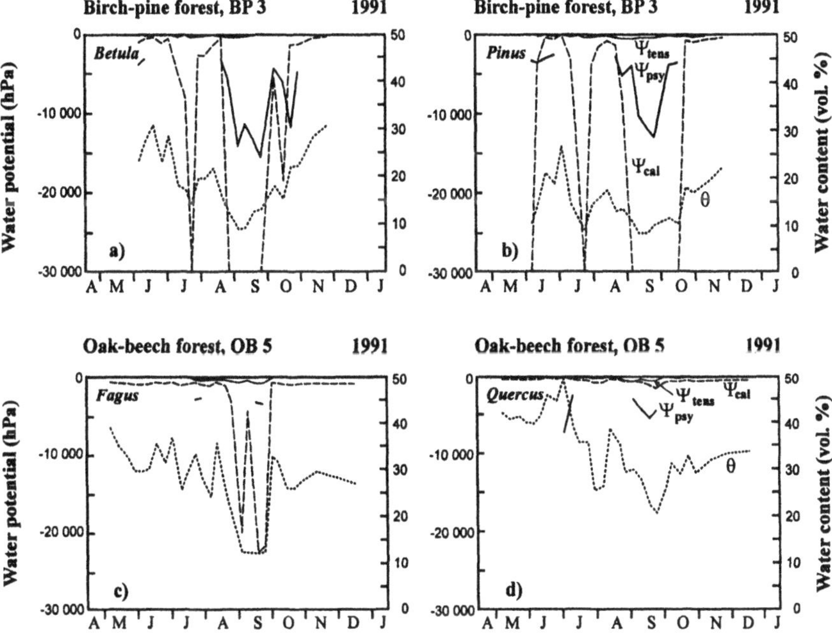

Fig. 1-27. Seasonal course of matric potential (Ψ_m) and volumetric water content (θ) in the organic Ofh-layers of the birch-pine forest BP3 (forest floor under *Betula* trees: (a), or *Pinus* trees: (b)) and oak-beech forest OB5 (under *Fagus*: (c), or *Quercus*: (d)) over the vegetation period 1991. Given are tensiometer readings in the mineral topsoil (2 cm below organic Oh horizon, Ψ_{tens}), values of soil psychrometers in the Ofh horizon (Ψ_{psy}) and Ψ values calculated from the θ/Ψ_m relationship established by laboratory desorption (Ψ_{cal}). θ stands for the volumetric water content of the Ofh-layers as determined gravimetrically.

The three methods for measuring the soil water potential in organic material gave highly different results in 1991 with the psychrometer values being significantly lower than the tensiometer readings. The values calculated from θ/Ψ_m curves showed a high temporal variability with very low potentials measured during drought. From the psychrometer and the θ/Ψ_m curve data it appears that much lower matric potentials (in the range of -1.5 MPa) occur in the organic layers than in the sandy mineral soil during drought. Although the absolute values are not very reliable, the matric potential values in the organic horizons reflect a negative gradient in the sequence *Quercus-Fagus-Betula-Pinus* humus (Fig. 1-27). Thus, organic material under *Quercus* (Fig. 1-27d) revealed a much more favourable moisture regime than humus of the other tree species or heather.

1.5 Changes in ecosystem water turnover

1.5.1 Canopy interception, throughfall and stemflow

Canopy interception was significantly smaller in the heathland than in the two forest communities (25 vs. 30-31% of bulk precipitation during mid-summer, Table 1-9). This difference should result primarily from the small leaf area index of the heathland (1.7 m^2 m^{-2}, Leuschner and Rode 1999). On the other hand, interception was remarkably similar in the birch-pine and oak-beech forests despite a significantly higher leaf area index in the latter community (5.3 vs. 4.0 m^2 m^{-2}). Thus, factors other than leaf area must be responsible in this case including leaf size and orientation, and canopy structure.

Betula, Pinus, Quercus and *Fagus* trees differed markedly with respect to stemflow. Measurements from April to October, 1991, showed significantly higher stemflow rates for the late-successional *Fagus* trees than for the three other tree species ranging from 3.5% of bulk precipitation (*Fagus*) to 0.8% (*Quercus*, Table 1-10). This variability is a consequence of differences in branch angle, bark surface structure and leaf area index among the four tree species. No relationship was found between successional position of a tree species and its stemflow rate. The species differences in stemflow were smaller in periods of high rainfall (as in July 1991) when *Quercus* and *Fagus* showed comparable rates (5.1 and 6.3% of precipitation, respectively). These data show that stemflow depends not only on tree species but also on rainfall intensity and season (cf. Benecke 1984).

Table 1-9. Bulk precipitation, canopy interception and throughfall plus stemflow in the three successional communities during mid-summer (14 May - 17 Sep), the vegetation period (18 Apr - 29 Oct) and over the year 1991 (in mm; numbers in parentheses: proportion of bulk precipitation, in %). Average coefficients of variation (std/mean) of bulk precipitation data were 0.04 in all three communities, 0.14, 0.28 and 0.13 for throughfall data in heathland, birch-pine forest and oak-beech forest, respectively, and 0.28 and 0.32 for stemflow in birch-pine and oak-beech forest, respect.; for number of samplers see Table 2.

	14 May-17 Sep	18 Apr-29 Oct	1 Jan-31 Dec
	Bulk precipitation (mm)		
Calluna heathland	265 (*100*)	372 (*100*)	638 (*100*)
Birch-pine forest	265 (*100*)	372 (*100*)	638 (*100*)
Oak-beech forest	253 (*100*)	372 (*100*)	669 (*100*)
	Canopy interception (mm)		
Calluna heathland	66 (*25*)	88 (*24*)	141 (*22*)
Birch-pine forest	80 (*30*)	120 (*32*)	198 (*31*)
Oak-beech forest	78 (*31*)	120 (*32*)	187 (*28*)
	Throughfall and stemflow (mm)		
Calluna heathland	199 (*75*)	284 (*77*)	497 (*78*)
Birch-pine forest	185 (*70*)	252 (*68*)	440 (*69*)
Oak-beech forest	175 (*69*)	252 (*68*)	482 (*72*)

1.5.2 Stand evapotranspiration

In the vegetation period of 1991, daily totals of stand evapotranspiration (ET, which include transpiration, soil evaporation and evaporation of intercepted water on rainy days) typically ranged between 1 and 2 mm in the *Calluna* heathland (with maxima as high as 4 mm), and between 2 and 4 mm in the two forest communities (with maxima reaching 5.5 mm in a high-radiation period in early July, 1991, Figs. 1-28 and 1-29). The ET rates of the two forest communities exceeded the corresponding heathland value by 42 to 46% in the period April to October, 1991. This large difference between heathland and forest communities is, in part, the consequence of a more than 25% higher rainfall interception in the two forests

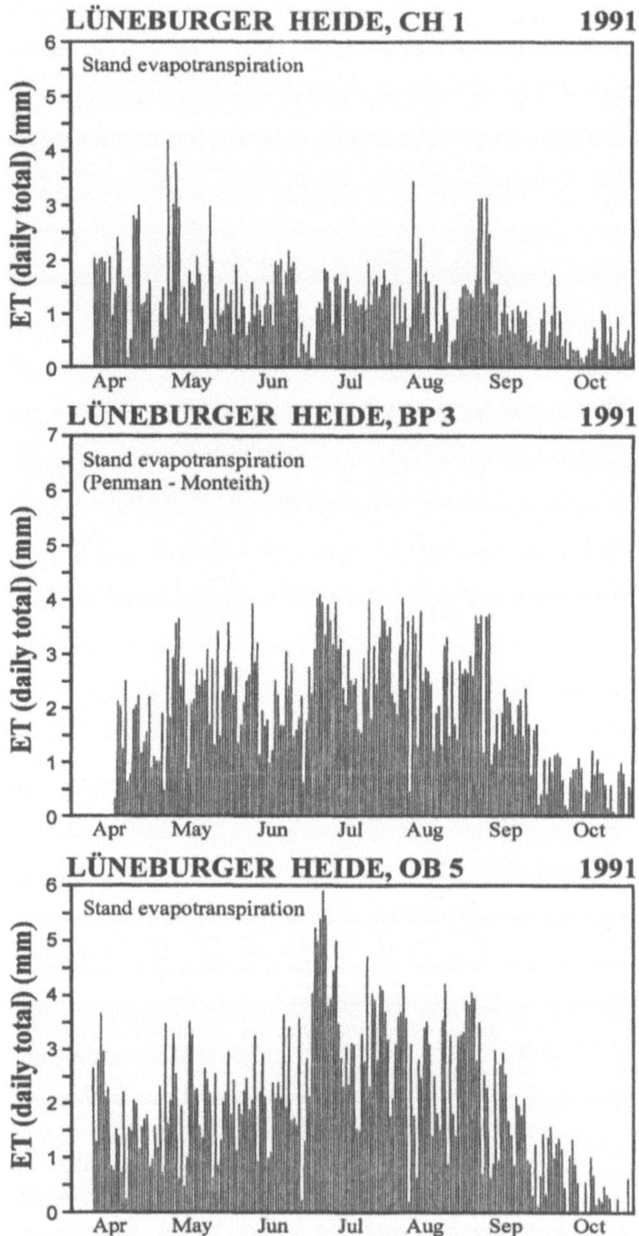

Fig. 1-28. Daily totals of evapotranspiration of the heathland (CH1), birch-pine forest (BP3) and oak-beech forest site (OB5) over the vegetation period 1991 (according to energy balance (Bowen ratio) measurements at CH1 and OB5, and calculations by the Penman-Monteith equation (big leaf formulation) at BP3.

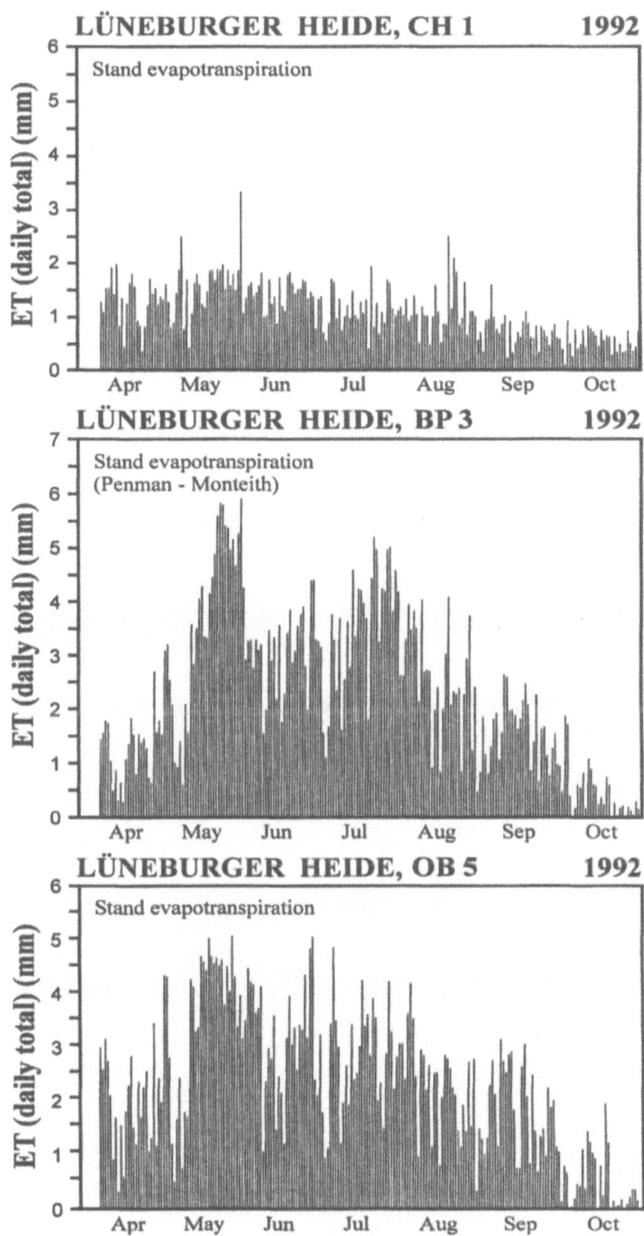

Fig. 1-29. Daily totals of evapotranspiration of the heathland (CH1), birch-pine forest (BP3) and oak-beech forest site (OB5) over the vegetation period 1992 (according to energy balance (Bowen ratio) measurements at CH1 and OB5, and calculations by the Penman-Monteith equation (big leaf formulation) at BP3.

Table 1-10. Stemflow of *Betula*, *Pinus*, *Quercus* and *Fagus* trees during summer 1991 in absolute (mm) and relative values (% of bulk precipitation, sites BP3 and OB5). Each 4 birches and pines, 13 beeches and 3 oaks were sampled, and the data extrapolated to mono-specific stands with typical stem densities.

	10 Apr-31 Oct		1 Jul-31 Jul	
	(mm)	(%)	(mm)	(%)
Betula	4.9	1.3	2.6	3.1
Pinus	9.0	2.4	4.4	5.3
Quercus	3.0	0.8	4.2	5.1
Fagus	13.3	3.5	6.3	6.3

Table 1-11. Totals of canopy rainfall interception, canopy transpiration and soil (litter) evaporation (in mm) in the period 1 April - 31 October, 1991, of the three successional communities (sites CH1, BP3 and OB5, respectively). Interception is calculated as the difference of bulk precipitation and canopy throughfall plus stemflow; canopy transpiration data refer to energy balance calculations (minus interception and soil evaporation values) in the case of heathland and oak-beech forest, and to calculations with the Penman-Monteith equation in the case of the birch-pine forest; soil evaporation was estimated from simplified energy balance calculations at the forest floor. Interception (%): Interception in % of stand evapotranspiration. n.d. - not determined.

	Calluna heathland	Birch-pine forest	Oak-beech forest
Canopy interception	96	123	122
Canopy transpiration	170	321	288
Soil evaporation	n.d.	51	44
Stand evapotranspiration	266	495	454
Interception (%)	36	25	27

(see Table 1-9) which significantly contributed to stand evapotranspiration. However, the main reason of higher ET rates in the forest stands are elevated canopy transpiration rates. In the vegetation period of 1991, calculated transpiration totals were 40 to 50% higher in the birch-pine and the oak-beech forests than in the heathland (Tab. 1-11). Thus, the initial heathland community differs from the two forest communities with respect to the evapotranspiration losses twofold, (i) by a markedly smaller height of ET, and (ii) by a larger proportion of interception in ET (36 vs. 25 or 27%, Table 1-11).

A comparison between the pine-dominated pioneer forest and the beech-domi-
nated late-successional forest community shows that the former had a slightly (8%)
higher stand evapotranspiration which results from both a higher canopy transpiration and an elevated soil evaporation loss (Table 1-11). The difference, however, is
small and might well be the consequence of measurement errors in ET, or the fact
that two different methods (Bowen ratio energy balance calculations vs. application of the Penman-Monteith equation, see Table 1-2) for estimating ET were used.
On the other hand, our radiation measurements gave roughly 9% higher net radiation inputs for the birch-pine forest in mid-summer 1991 as compared to the oak-
beech stand (Leuschner and Rode 1999) which primarily is the consequence of
different short-wave reflectivities. A higher radiation load should in fact result in
an elevated evapotranspiration rate in this conifer-dominated pioneer forest community.

1.5.3 Water turnover in the organic layers

For the summer months (May - September) of 1991 and 1992, the Figures 1-30 and
1-31 give the results of water balance calculations for the organic layers in the
pioneer and the late-successional forest communities. Based on daily canopy
throughfall data, the Forest Floor Water Flux Model gave daily rates of all component fluxes of the water balance equation. They are shown as monthly averages in
the graphs.

When contrasting throughfall with *seepage* rates in the two forest communities,
it is evident that, during summer, only wet months (such as June 1991 and August
1992) showed a significant percolation through the organic layers that led to infiltration into the mineral soil. During the summers of 1991 and 1992, only 60% of
all throughfall events resulted in drainage out of the organic profile in the oak-
beech forest. Moreover, only 56% (1991) or 37% (1992) of the throughfall water
reached the mineral soil (Leuschner 1998). In the birch-pine forest, 54 (1991) or
59% (1992) of the throughfall water infiltrated into the mineral soil. Thus, the
organic layers of the pioneer and late-successional forest communities effectively
buffered the mineral soil from large amounts of summer rainfall.

The Forest Floor Water Flux Model calculated remarkably constant *water uptake* rates of 0.5 mm d^{-1} for the tree roots in the organic layers of the oak-beech
forest. During mid-summer (August 1991 and August 1992), values peaked at 0.8
and 1.0 mm d^{-1} (Figure 1-31). According to the model, roots showed high uptake
rates in the organic layers even in dry July 1991. This result is consistent with data
on the forest floor water content in July 1991 that indicate water reserves in height
of 18 to 35 mm in the organic layers of this stand (Fig. 1-15). During May to September, 1991, about half of the throughfall water that entered the organic layers
was extracted by tree roots in this horizon.

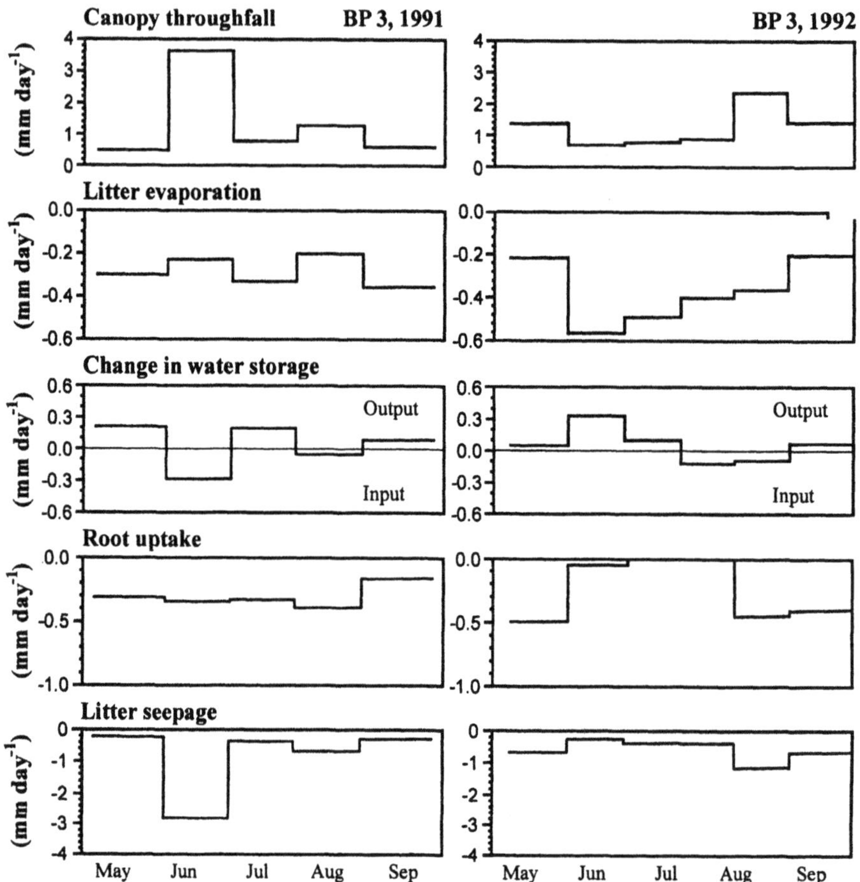

Fig. 1-30. Monthly averages of the main water fluxes in the organic profile (L- and O_{fh}-layers) of the birch-pine forest in summer 1991 (left) and 1992 (right) (results from calculations with the Forest Floor Water Flux Model, Leuschner 1998).

Given the small depth of the organic profile (92 mm), root water uptake was remarkably high in the forest floor of the late-successional community with 88 and 89 mm in the summers 1991 and 1992, respectively. This indicates a very rapid water turnover in the organic layers. Root water uptake in the organic layers of the birch-pine forest was less important and approached zero during the dry summer months of 1992 (Fig. 1-30). This matches with lower mean water contents and a smaller fine root biomass in the forest floor of the pioneer forest compared to the late-successional forest.

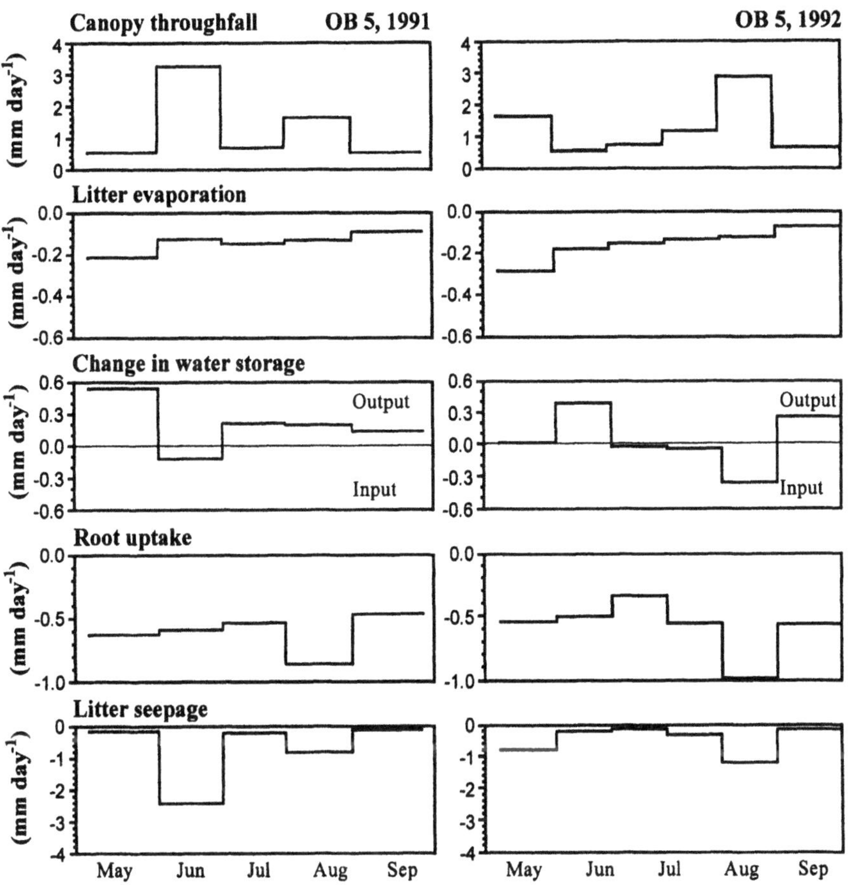

Fig. 1-31. Monthly averages of the main water fluxes in the organic profile (L- and Ofh-layers) of the oak-beech forest in summer 1991 (left) and 1992 (right) (results from calculations with the Forest Floor Water Flux Model, Leuschner 1998).

Litter evaporation as estimated from both energy balance calculations at the forest floor and gravimetric water loss determination showed maximum rates of 0.2 mm d^{-1} during the vegetation period and 0.3 mm d^{-1} in the leafless season (e.g. April 1992) for the oak-beech forest (Fig. 1-31). Rates were higher in the birch-pine forest (0.3 to 0.5 mm d^{-1}) due to a higher transmissivity of net radiation to the forest floor (Leuschner and Rode 1999).

1.5.4 Drainage and ecosystem water balance

Water that infiltrates into the soil and is not taken up by roots eventually may leave the rooted soil volume as drainage water. Vegetation types with a higher evapotranspiration loss are likely to have a lower drainage and should be less effective with respect to the recharge of ground water reserves. Due to its comparably low transpiration and interception rates, the heathland community had a more than three times higher annual drainage in 1991 than the late-successional oak-beech forest (332 and 100 mm, respectively, Table 1-12). Moreover, drainage was observed in the former community not only in the cold winter months but also during several wet summer periods and comprised more than 50% of bulk precipitation.

Table 1-12. Water fluxes (in mm) in 1991 in heathland (CH1), pioneer birch-pine forest (BP3) and late-successional oak-beech forest (OB5). Interception was calculated as the difference between bulk precipitation and throughfall plus stemflow, canopy transpiration and soil evaporation were derived from Bowen ratio energy balance measurements (CH1 and OB5) or calculations with the Penman-Monteith equation (BP3). Changes in soil water storage (0-70 cm depth) were received from soil water content measurements by gravimetry or tdr. Drainage was obtained by solving the water balance equation.

	Calluna heathland	Birch-pine forest	Oak-beech forest
Bulk precipitation	638	638	669
Interception	141	198	187
Transpiration	195	325	288
Soil evaporation	0[1]	65	64
Change in soil water storage[2]	-30	+35	+30
Drainage at 70 cm depth	332	15	100

[1]Included in transpiration value, [2]Jan 1-Dec 31, 1991

Using a one-dimensional soil water flux model, 0.5 to 0.8 mm were calculated as typical daily drainage rates during the wet winter months in the deciduous oak-beech forest (Fig. 1-32: upper panel, data from Hölscher, unpubl.). This rate rapidly declined with leaf expansion in May from day 120 onwards.

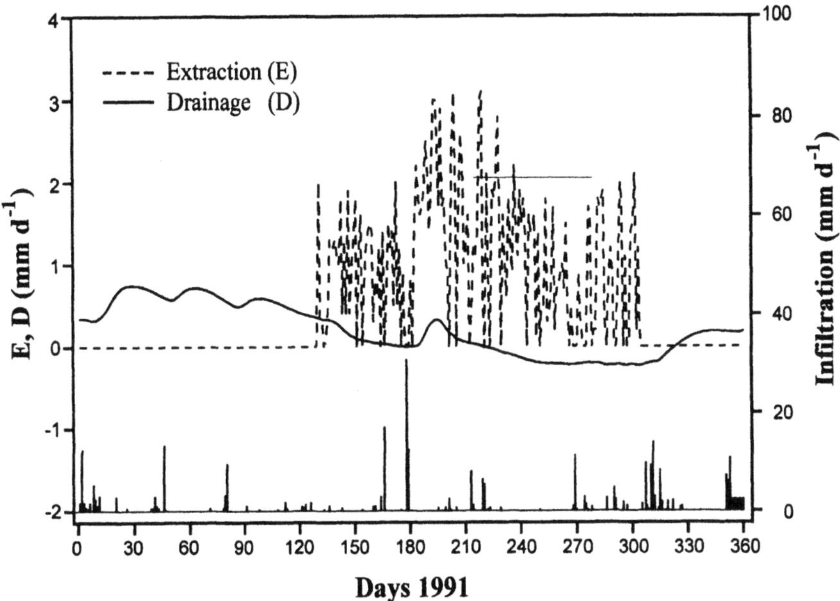

Fig. 1-32. (Upper panel) Daily totals of water extracted by roots in the mineral soil (E) and of water drained out of the soil profiles (D) under the late-successional oak-beech forest in the year 1991 (E was derived from Bowen ratio-energy balance measurements above the canopy, D was received from simulations with a numerical one-dimensional soil water flux model based on Darcy's law). Negative drainage values indicate capillary rise. (Lower panel) Infiltration rates (mm d^{-1}) into the mineral soil surface as calculated from canopy throughfall measurements and simulations with a forest floor water flux model (see Leuschner 1998).

Between July and November, 1991, virtually no drainage occurred in this stand due to the upward-directed water potential gradient in the soil profile, and soil matric potentials remained at constantly low values over the summer period (Fig. 1-32). In the pine-dominated pioneer forest, a zero-drainage period was observed not only during summer but it also extended into the first half of the winter. This must be seen as a result of both higher interception losses in winter and slightly higher transpiration rates in summer (see Table 1-12). As a result, annual drainage was estimated at only 15 mm (or 2% of bulk precipitation) for this community in 1991 which is by far the lowest rate among the three successional communities.

1.6 Discussion

Forest succession and soil moisture regime

Vegetation change with succession should alter the soil moisture regime less mark-edly than it changes water fluxes. A comparison of the soil moisture regimes under heathland, pioneer and late-successional forest in the Lüneburger Heide indeed shows that the three communities had rather similar soil moisture regimes despite large differences in canopy structure and phenology. However, important differ-ences in soil hydrology existed with respect to (i) the water storage in the organic layers, and (ii) the water content of the mineral topsoil which was significantly higher under the oak-beech forest compared to the two other communities.

The accumulation of organic material at the forest floor resulted in a consider-able increase of the humus water storage capacity when proceeding from the heath-land to the late-successional stage. The relative contribution of the organic layers to the soil water reserves rose from negligible values in the heathland with thin Xeromors to about 27% of the profile total in the oak-beech forest community with deep Hemihumimors. This reflects a significant improvement of plant water avail-ability not only in wet but also in dry seasons. Moreover, the organic layers con-tributed increasingly to the stand water demand. At least in the late-successional community, the organic layers were an important source for plant water uptake with 37% of the stand water consumption provided by this compartment (Leuschner 2001). Higher moisture contents during summer in late stages of suc-cession also must favour decomposition, humification and root growth. Thus, hu-mus accumulation in forest successions on poor soils may be a key process not only for nutrient but also for water supply to plants.

Higher water contents in the mineral topsoil of the late-successional community compared to the other two vegetation types are a surprising result since plant water uptake must be particularly high in this compartment due to the very high fine root densities found in the topsoil of the oak-beech forest (Leuschner 2001). Moreover, high water contents contradict the view that higher transpiration and interception rates in forest ecosystems should lead to lowered soil moisture compared to grass-land or heathland. Indeed, Sharma et al. (1987) observed an increase in soil water storage with the conversion of forest to pasture in Australia. Similarly, Hodnett et al. (1996) measured higher soil water storages under pasture than under forest in Amazonia during periods of drought, which reflects the higher evapotranspiration of the forest.

Several factors may have contributed to the unexpected soil moisture effect un-der the oak-beech forest in this study. Among them are (i) particularly low soil dry mass densities and corresponding high soil porosities that allow higher maximum water contents, (ii) a sheltering effect of the thick and moist organic layers in this community that reduces the evaporation from the mineral soil, (iii) energy limitation of forest floor evaporation due to a low net radiation transmissivity, and (iv) the possibility of significant hydraulic lift by the root system of *Fagus* and *Quercus* trees which would transport water from deep horizons to the organic layers with highest root activity. Additional research is needed to answer this question.

The sandy soil profiles in the Lüneburger Heide showed a number of hydrologic characteristics which are important for assessing soil water availability in the communities of the heathland-to-forest succession and that contradict current views on the hydrology of sandy soils:

1) Clearly, the field capacity of the sandy soils cannot be defined as the water content at -100 (or -300) hPa but has to include water in the range of about -20 to -100 hPa as well. In sandy soil profiles under two pine forests in Lower Saxony (Germany), Simorangkir (1994) found matric potential values during winter that were also in the range of -20 to -60 hPa. These results indicate higher field capacities as well.

2) As a consequence, the storage capacity of plant-available water (PW) in these soils must be much higher than previously thought. The recent study shows that maximum water reserves of PW as high as 150 to 170 mm per 1 m profile may exist during winter under the three communities. In contrast, Brechtel and van Hoyningen-Huene (1979) assumed only 70 mm, and Renger and Strebel (1980) 92 mm of plant-available water per 1 m profile for medium-grained sand. Particularly in early summer periods, sandy diluvial soils might be less dry than previously thought.

3) Soil profiles under all three communities showed a remarkable anomaly in their water content and matric potential patterns with minima occurring at 40 to 60 cm depth. These minima do not correspond to maxima of fine root density (d < 2 mm diameter) that was much higher in the wetter mineral topsoil. However, coarse roots (2 < d < 5 mm) of *Quercus* and *Fagus* showed their maximum density at this depth; this poses the question on the role of tree coarse roots in water uptake.

4) During summer (May/June to October), virtually no drainage occurred under the two forest communities as a consequence of an upward-directed water potential gradient in the soil. In addition, thick organic layers buffered the mineral soil from a large proportion of the summer throughfall. Thus, from June to October, water flow in the sandy soils is dominated by a slow and steady upward movement in the direction of the topsoil where high fine root densities were found.

Forest succession and water fluxes

Plant surfaces are key determinants of the water turnover in forest ecosystems. Three fluxes of the ecosystem water cycle are especially sensitive to changes in leaf and fine root surface areas, (i) transpiration, (ii) root water uptake, and (iii) rainfall interception (Aston 1979, Landsberg and McMurtrie 1984, Kelliher et al. 1995). Leaf area was found to be positively correlated with stand transpiration in a number of woody and herbaceous plant communities (e.g. Grier and Running

1977, Waterloo et al. 1999, Köstner et al. 2001). Similarly, higher interception losses are to be expected in stands with larger leaf areas since the average rainfall storage capacities of forest stands are influenced by the leaf area index (Rutter and Morton 1977). Other processes in the ecosystem water cycle such as stemflow, soil evaporation, soil water storage, runoff and drainage are less dependent on plant surface areas and are likely to be influenced by vegetation structure more indirectly.

Long-term secondary and primary succession may alter vegetation structure fundamentally. Heathland-to-forest succession in the Lüneburger Heide led to large changes in both stand leaf area index (from 1.7 to 5.3 m^2 m^{-2}) and vegetation height (from 0.2 m to 30 m) between early and late stages over a time span of about 300 years (Leuschner and Rode 1999). Thus, above-ground surfaces increased about threefold, vegetation height even by a factor of 150 during this sequence. Changes in below-ground surfaces are much more difficult to measure and data on successional changes in fine root surface areas (or number of fine root tips or ectomycorrhizas) are lacking in the literature. For the studied succession, data on fine root biomass of heathland, birch-pine forest and oak-beech forest show an only slight increase in fine root biomass (diameter < 2 mm) over succession (from 430 to 566 g m^{-2}). Apparently, no increase in fine root surface area occurred because *Calluna* fine roots have a much higher specific surface area than roots of late-successional trees with comparably thick fine roots (Leuschner and Rode 1999, Leuschner 2001). However, the functional types of mycorrhizal fungi markedly change from ericoid to ectomycorrhizal during this succession. This should influence the active below-ground surface area of the community and could also affect the water uptake capacity of the successional plants. It is concluded that, while a clear increase in above-ground evaporating plant surfaces exists, no such trend seems to occur in the below-ground water absorbing surfaces during this succession. The failure to identify directional change in root surfaces is partially due to inadequate data on mycorrhizal surfaces and their contribution to plant water uptake.

The changes in canopy interception, throughfall and stemflow were remarkably small over heathland-to-forest succession given the fact that leaf area index, canopy structure and vegetation height are fundamentally different between early and late stages of this succession. Only by 25 to 29% higher interception losses were found in the two forest communities compared to the heathland despite a two to three times smaller leaf area index in the latter. It appears that effects of a changing leaf area on interception and throughfall are, to a certain degree, balanced by effects of structural changes in the vegetation. The latter include a shift from predominantly evergreen (*Calluna* and *Pinus*) to summer-green (*Fagus* and *Quercus*) vegetation types, and changes in the species-specific canopy surface and branching structure, and bark surface properties among the dominant early or late-successional species of this succession. Storage capacities for intercepted water were found to be different for hardwoods (about 1 mm) and conifers (about 2 mm per unit leaf area, Zinke 1967). This difference might explain the high interception in the pine and heather canopies despite a relatively small leaf area index.

Evapotranspiration is a major output term in the ecosystem water budget. It is partly under plant control through the processes of canopy leaf area expansion and stomatal function which both may change over succession. However, successional alterations in physical factors that control the evaporation process might even be more influential: vegetation height increases during heathland-to-forest succession by a factor of 150. This size effect is likely to improve the aerodynamic coupling between canopy and atmosphere (McNaughton and Jarvis 1983) and, from this physical perspective, higher fluxes of water vapour into the atmosphere should be expected in tall late-successional compared to low early-successional communities. However, a review of micrometeorological flux measurements by McNaughton and Jarvis (1983) showed that the transpiration rate from forest was generally less than that from grassland or arable crops, whereas the evaporation rate of intercepted water was usually somewhat higher. Thus, whether total evapotranspiration losses (i.e. transpiration plus interception) are higher over forest or non-woody vegetation, should depend primarily on the ratio of transpiration to evaporation of intercepted water. It may be noted, however, that the majority of the studies analysed by McNaughton and Jarvis (1983) were conducted in oceanic climates with ample moisture supply and only limited radiation input.

Further indirect information on changes in evapotranspiration due to alterations in vegetation cover and/or structure is provided by runoff studies in paired clearcut and forested catchments. Hibbert (1967), Bosch and Hewlett (1982), Greenwood (1992) and Bruijnzeel (1996) concluded from a large number of investigations in both temperate and tropical regions that deforestation increases water yield and reforestation decreases it. Furthermore, there are a number of studies that investigated the reduction in runoff by reforestation, indicating increases in interception and transpiration with the planting of forest on grassland or arable fields (e.g. Trimble et al. 1987, van Wyk 1987). From soil moisture or lysimeter studies Metz and Douglass (1959), Douglass (1967) and Käppeli and Schulin (1988) concluded that forest used more water and water from deeper depths than the shallow rooted grass and, thus, drainage was always greater under grass than under hardwoods. Similarly, Newson and Calder (1989) compared forested and grassland catchments in upland UK and calculated annual ET rates that were double as high for forest as for grassland. Although ET may be similar for grass and forest vegetation under very humid conditions with readily available moisture (e.g. McNaughton and Jarvis 1983), the bulk of evidence from landscape or patch scale studies shows that low-statured vegetation uses less water than forest. Thus, in most cases, succession from heather, grassland or arable fields to forest should lead to significant increases in evapotranspiration and to corresponding decreases in drainage and/or runoff. The data of this study principally support this view. The decline in streamflow to pre-disturbance levels with proceeding secondary succession follows canopy closure and may depend on climate and plant growth rates: the process can can last for decades in temperate regions (Hibbert 1967), but may proceed much faster in the tropics with a rapid regrowth of woody vegetation (Bruijnzeel 1996).

Three causes may be responsible for the observed increase in evapotranspiration, (i) a higher leaf area index (or leaf area duration) of woody vegetation, (ii) a more limited water availability to the shallow-rooted non-woody vegetation, and (iii) higher losses of intercepted (or transpired) water due to a better aerodynamic coupling of woody vegetation. Apparently, all three factors contributed to the observation of a by 40-45% higher evapotranspiration rate of pioneer and late-successional forests compared to the heathland: the leaf area index showed a threefold increase, soil water depletion proceeded to increasingly greater depths, and evaporation rates of intercepted water increased by 25 to 29% between early and late stages of succession.

The evapotranspiration rates measured for the *Calluna* heathland in this study were remarkably small (266 mm during April-October 1991). In the UK, heathland was found to have a smaller evapotranspiration than grassland if annual rainfall was < 1250 mm. In the Lüneburger Heide, periodic summer drought may have contributed to particularly low transpiration rates. The annual evapotranspiration rate determined in 1991 for the birch-pine forest (588 mm with 198 mm referring to interception, 325 mm to transpiration, and 65 mm to soil evaporation) compares well with ET rates calculated from Bowen ratio and soil hydrological data of the 40-yr-old Fuhrberg pine forest (40 km south of site BP3, 670 or 614 mm evapotranspiration, and 383 or 373 mm transpiration in 1984 or 1985, respectively, Simorangkir 1994).

Increasing evapotranspiration losses during succession should correspond to decreasing drainage rates. The annual water balance data for the three successional communities show that, if heathland-to-forest succession proceeds in a large area, the replacement of heathland by pioneer forest must reduce drainage greatly and, thus, will negatively influence groundwater recharge on a landscape scale. The observed very small annual drainage rates under the birch-pine forest (15 mm) match with results of Brechtel and von Hoyningen-Huene (1979) who used a rough input-output analysis to calculate 41 mm as annual drainage for *Pinus sylvestris* forests of 40-80 years, and zero drainage for stands > 80 years in the Rhine valley of Central Germany. Based on Bowen ratio and soil hydrological data, Simorangkir (1994) obtained annual drainage rates of 129 mm for the Fuhrberg pine forest in the moderate dry year 1984, and of -51 mm (i.e. a negative soil water balance) in the very dry year 1985. Both sites received about 660 mm of annual rainfall.

In later stages of the Lüneburger Heide succession with dominance of deciduous trees, drainage significantly increases again according to the data obtained. This is in accordance with Brechtel and von Hoyningen-Huene's (1979) data who reported 91 and 113 mm annual drainage under beech forests of 40-80 years or > 80 years.

They also found significantly higher rates under beech than under nearby pine forests. Drainage is not the only parameter that passes through a minimum in the pioneer forest stage during the heathland-to-forest succession. The ratio of soil water availability and water demand (or uptake) rapidly decreases when proceeding from the heathland to the pioneer forest stage. This indicates that water limitation of plant growth must be most severe at the pioneer forest stage.

Acknowledgements

This investigation was part of a comprehensive study on the mechanisms of heathland-to-forest succession, which included research at both the ecosystem and the plant ecophysiological levels. It was conducted at the Department of Plant Ecology, University of Göttingen, in collaboration with Forschungszentrum Waldökosysteme, University of Göttingen. Financial support was obtained from the Federal Ministry of Research and Technology (project No. P.6.3.8, FZW, and contract No. 0339251A), and EC contract No. EV4V-0148-C(BA)) which is gratefully acknowledged. This research could not have been conducted without the support by Katharina Backes, Andrea Dageförde, Gabi Görlitz, Michael Rode and Jochen Schenk, who collected part of the field data, and Bernd Raufeisen, who prepared much of the drawings. I thank them all. I am also very grateful to M. Runge who offered a working niche for this succession project in the institute.

References

Aston AR (1979) Rainfall interception by eight small trees. J Hydrol 42:383-396

Backes K, Leuschner C (2000) Leaf water relations of competitive *Fagus sylvatica* L. and *Quercus petraea* (Matt.) Liebl. trees during four years differing in soil drought. Can J For Res 30:335-346

Baldocchi DD, Hicks BB, Myers TP (1988) Measuring biosphere-atmosphere exchanges of biologically related gases with micrometeorological methods. Ecology 69:1331-1340

Bazzaz FA, Sipe TW (1987) Physiological ecology, disturbance, and ecosystem recovery. In: Schulze ED, Zwölfer H (eds) Potentials and Limitations of Ecosystem Analysis. Ecol Stud 61. Springer, Berlin Heidelberg:203-227

Benecke P (1984) Der Wasserumsatz eines Buchen- und eines Fichtenwaldökosystems im Hochsolling. Schr Forstl Fak Univ Göttingen 77:1-158

Berendse F (1990) Organic matter accumulation and nitrogen mineralization during secondary succession in heathland ecosystems. J Ecol 78:413-427

Bormann BT, Sidle RC (1990) Changes in productivity and distribution of nutrients in a chronosequence at Glacier Bay National Park, Alaska. J Ecol 78:561-578

Bosch JM, Hewlett JD (1982) A review of catchment experiments to determine the effects of vegetation changes on water yield and evapotranspiration. J Hydrol 55:3-23

Brechtel HM, von Hoyningen-Huene J (1979) Einfluß der Verdunstung verschiedener Vegetationsdecken auf den Gebietswasserhaushalt. Schriftenreihe DVWK (Braunschweig) 40:172-231

Bruijnzeel LA (1996) Predicting the hydrological impacts of land cover transformation in the humid tropics: the need for integrated research. In: Gash JHC, Nobre CA, Roberts JM, Victoria RL (eds) Amazonian Deforestation and Climate. J Wiley & Sons, Chichester:15-25

Brutsaert W (1982) Evaporation into The Atmosphere. Reidel, Dordrecht

Calder IR (1990) Evaporation in The Uplands. Wiley, Chichester

Christensen NL, Peet RK (1981) Secondary forest succession on the North Carolina piedmont. In: West DC, Shugart HH, Botkin DB (eds) Forest Succession: Concepts and Applications Springer, New York:230-245

Douglass JE (1967) Effects of species and arrangement of forests on evapotranspiration. In: Sopper WE, Lull HW (eds) Forest Hydrology. Pergamon, Oxford:451-461

Ehlers W (1991) Leaf area and transpiration efficiency during different growth stages in oats. J Agric Sci 116:183-190

Emmer IM (1995) Humus form and soil development during a primary succession of monoculture *Pinus sylvestris* forests on poor sandy substrates. PhD thesis, Univ of Amsterdam

Flüggen C (1991) Die Evaporation von Kiefern unter Berücksichtigung des Grundwasserabstandes. Ber Inst Meteor Klimatol Univ Hannover 39:1-82

Gerlach A, Albers EA, Broedlin W (1994) Development of the nitrogen cycle in the soil of a coastal dune succession. Acta Bot Neerl 43:189-203

Goldberg DE (1990) Components of resource competition in plant communities. In: Grace JB, Tilman D (eds) Perspectives on Plant Competition. Acad Press, San Diego:27-49

Greenwood EAN (1992) Deforestation, revegetation, water balance and climate: an optimistic path through the plausible, impractible and controversial. Adv Bioclimatol 1:89-154

Grier CC, Running SW (1977) Leaf area of mature northwestern coniferous forests: relation to site water balance. Ecology 58:893-899

Griese F (1987) Untersuchungen über die natürliche Wiederbewaldung von Heideflächen im niedersächsischen Flachland. PhD thesis, Univ of Göttingen

Heinken T (1995) Naturnahe Laub- und Nadelwälder grundwasserferner Standorte im niedersächsischen Tiefland: Gliederung, Standortsbedingungen, Dynamik. Dissert Bot 239:1-311

Hertel D (1999) Das Feinwurzelsystem von Rein- und Mischbeständen der Rotbuche: Struktur, Dynamik und interspezifische Konkurrenz. Dissert Bot 317:1-190

Hibbert AR (1967) Forest treatment effects on water yield. In: Sopper WE, Lull HW (eds) Forest Hydrology. Pergamon, Oxford:527-543

Hodnett MG, Oyama MD, Tomasella J, Marques Filho A de O (1996) Comparisons of long-term soil water storage behaviour under pasture and forest in three areas of Amazonia. In: Gash JHC, Nobre CA, Roberts JM, Victoria RL (eds) Amazonian Deforestation and Climate. J Wiley & Sons, Chichester:57-78

Hölzer R (1982) Wasserhaushaltsuntersuchungen der Streu- und obersten Bodenschicht eines Fichtenbestandes unter Verwendung von Modellrechnungen. Beiträge zur Hydrologie (Kirchzarten), Sonderheft 4:117-144

Kelliher FM, Leuning R, Raupach MR, Schulze E-D (1995) Maximum conductances for evaporation from global vegetation types. Agric For Meteor 73:1-16

Käppeli T, Schulin R (1988) Lysimeteruntersuchungen zur Wasserbilanz von Pappel, Weisserle, Fichte und Gras auf einem sandigen Boden über Schotter. Schweiz Z Forstwes 139:129-143

Kellman M, Roulet N (1990) Nutrient flux and retention in a tropical sand-dune succession. J Ecol 78:664-676

Klinka K, Green RN, Trowbridge RL (1993) Towards a taxonomic classification of humus forms. For Sci Monogr 29

Köstner B, Alsheimer M, Wedler M, Scharfenberg H-J, Zimmermann R, Falge E, Joss U, Tenhunen JD (2001) Controls on evapotranspiration in spruce forest stands. In Tenhunen JD, Lenz R, Hantschel R (eds.) Ecosystem approaches to landscape management in Central Europe. Ecol Studies 147. Springer, Berlin:377-416

Landsberg JJ, McMurtrie R (1984) Water use by isolated trees. Agric Water Manage 8:223-242

Leuschner C (1993) Patterns of soil water depletion under coexisting oak and beech trees in a mixed stand. Phytocoenologia 23:19-33

Leuschner C (1994) Walddynamik auf Sandböden in der Lüneburger Heide, NW Deutschland. Phytocoenologia 22:289-324

Leuschner C (1998) Water extraction by tree fine roots in the forest floor of a temperate *Fagus-Quercus* forest. Ann Sci Forest 55:141-157

Leuschner C, Rode MW (1999) The role of plant resources in forest succession: changes in radiation, water and nutrient fluxes, and plant productivity over a 300-yr-long chronosequence in NW Germany. Perspect Plant Ecol Evol Syst 2:103-147

Leuschner C (2001) Changes in forest ecosystem function with succession in the Lüneburger Heide. In Tenhunen JD, Lenz R, Hantschel R (eds.) Ecosystem approaches to landscape management in Central Europe. Ecol Studies 147. Springer, Berlin:517-570

Leuschner C, Rode MW, Danner E, Lübbe K, Clauss C, Margraf S, Runge M (1993) Soil profile alteration and humus accumulation during heathland-forest succession in NW Germany. Scripta Geobot 21:73-84

McNaughton KG, Jarvis PG (1983) Predicting effects of vegtation changes on transpiration and evaporation. In: Kozlowski TT (ed) Water Deficits and Plant Growth, Vol VII. Academic Press, New York:1-47

Metz LJ, Douglass JE (1959) Soil moisture depletion under several Piedmont cover types. U.S. Dept Agr Tech Bull 1207:1-23

Miles J (1979) Vegetation Dynamics. Chapman and Hall, London

Newson MD, Calder IR (1989) Forests and water resources: problems of prediction on a regional scale. Phil Trans R Soc Lond B324:283-298

Renger M, Strebel O (1980) Beregnungsbedarf landwirtschaftlicher Kulturen in Abhängigkeit vom Boden. Wasser und Boden 32:572-575

Rode MW (1995) Aboveground nutrient cycling and forest development on poor sandy soil. Plant Soil 168-169:337-343

Rode MW (1999) Influence of forest growth on former heathland on nutrient input and its consequences for nutrition and management of heath and forest. For Ecol Manage 114:31-43

Rode MW, Schmitt U (1995) Nutrient distribution and enrichment within the above-ground biomass of three successional ecosystems. Aarhus Geoscience 4:45-52

Rode MW, Leuschner C, Clauss C, Gerdelmann V, Margraf S, Runge M (1993) Changes in nutrient availability and nutrient turnover during heathland-forest succession in NW Germany. Scripta Geobot 21:85-96

Rutter AJ, Morton AJ (1977) A predictive model of rainfall interception. III. Sensitivity of the model to stand parameters and meteorological variables. J Appl Ecol 14:567-588

Sharma ML, Barron RJW, Williamson DR (1987) Soil water dynamics of lateritic catchments as affected by forest clearing for pasture. J Hydrol 94:29-46

Simorangkir D (1994) Die Hysterese der pF-Kurve am Beispiel ihrer Auswirkung auf die Simulationsergebnisse für den Wasserhaushalt sandiger Kiefernwaldökosysteme. Ber Forschungsz Waldökosyst Univ Göttingen A119:1-130

Stewart BA, Nielsen DR (1990) Irrigation of Agricultural Crops. Series Agronomy No 30. Amer Soc of Agronomy, Madison, Wisconsin

Tilman D (1985) The resource ratio hypothesis of succession. Am Nat 125:827-852

Tilman D (1988) Plant Strategies and The Dynamics and Structure of Plant Communities. Monogr Pop Biol, Princeton Univ Press, Princeton, New Jersey

Trimble SW, Weirich FH, Hoag BL (1987) Reforestation and the reduction of water yield on the southern Piedmont since circa 1940. Water Resour Res 23:425-437

Viereck LA (1970) Forest succession and soil development adjacent to the Chena River in the interior Alaska. Arct Alp Res 2:1-26

Vitousek PM, Reiners WA (1975) Ecosystem succession and nutrient retention: a hypothesis. BioScience 25:376-381

Walker J, Thompson CH, Fergus JF, Tunstall BR (1981) Plant succession and soil development in coastal sand dunes of subtropical eastern Australia. West DC, Shugart HH, Botkin DB (eds) Forest Succession: Concepts and Applications. Springer, New York:107-131

Waring RH, Running SW (1998) Forest Ecosystems: Analysis at Multiple Scales. Academic Press, New York

Waterloo MJ, Bruijnzeel LA, Vugts HF (1999) Evaporation from *Pinus caribaea* plantations on former grassland soils under maritime tropical conditions. Water Resour Res 35:2133-2144

Wyk DB van (1987) Some effects of afforestation on stream flow in western Cape Province, South Africa. Water South Africa 13:31-36

Zinke PJ (1967) Forest interception studies in the United States. In: Sopper WE, Lull HW (eds) Forest Hydrology. Pergamon, Oxford:137-160

2 Environmental impacts on forest ecosystems

2.1 Manipulation of nutrient and water input of a Norway spruce ecosystem

A. Dohrenbusch*, M. Bredemeier, N. Lamersdorf***

*   Institute of Silviculture, Göttingen University, Büsgenweg 1,
    D-37077 Göttingen, Germany, e-mail: adohren@gwdg.de
** Forest Ecosystems Research Center, Büsgenweg 1,
    D-37077 Göttingen, Germany, e-mail: mbredem@gwdg.de
***Institute of Soil Science and Forest Nutrition, Büsgenweg 2,
    D-37077 Göttingen, Germany, e-mail: nlamers@gwdg.de

Abstract

In the mountainous region of a low mountain range (Solling hills) an ecosystem manipulation experiment with roof constructions underneath the canopy of a 60-year old Norway spruce stand is run since 1991. The experimental design of this ten-years lasting experiment is presented. The responses to artificially prepared, "pre-industrial" throughfall and to extended summer droughts with intensive rewetting are investigated in two parallel roof experiments and evaluated against a roof control and an ambient control plot. Hypotheses derived from forest ecosystem monitoring can thus be directly tested using this "laboratory in the field". The aim of this interdisciplinary experiment is to examine the long-term effects of a changed nutrient- and water-supply caused by a manipulated input. The results in terms of growth, vitality and nutrient conditions are presented in the following articles.

Key words: roof project, *Picea abies*, de-acidification, acid rain, clean rain, drought, experimental manipulation of nutrient and water input

2.1.1 Introduction and objectives

The effects of environmental parameters on forest ecosystems could theoretically be best investigated under laboratory conditions. Under laboratory conditions it is possible to modify single factors while others are kept constant. However, the transfer of the results thus obtained to the ecosystem is problematic. Conditions are required which allow the control of influence factors, but these conditions differ markedly from the natural conditions (for example root investigations in hydro-culture, photosynthesis measurements in growth chambers). In addition, results obtained under laboratory conditions do not give a realistic picture of the complex interactions in an ecosystem. Existing interrelationships and mutual dependencies can not be sufficiently considered. An alternative method is a long term observation of forest ecosystems under field conditions with parallel observations of the role of the environmental factors. The disadvantages of this method are the prolonged periods of observation required and the difficulty in determination of those parameters which have a strong effect on the ecosystems among a number of varying factors. In order to avoid these disadvantages, ecosystems as a whole or at least representative parts have to be exposed to controlled changes of the environment. This concept is the basis for the roof-project presented here. The objective of this interdisciplinary research project is to elucidate the following questions:

- What are the effects of a strongly reduced acid and nitrogen input on abiotic soil solutions and on the biotic part of the ecosystem?
- What is the effect of pollutant impacts in combination with climatic stress factors (drought)?
- Are damages due to immission and climate conditions reversible?
- What are the interactions between the sub-systems within the ecosystems?

The latter point is concerned with the relationships between the soil, the soil fauna and the above and below ground parts of the trees, carrying out analyses of the growth, and morphological and physiological studies. The obtained results should finally show whether environment oriented political measures have an effect on forest ecology.

The large scale experiment concentrates on two basic environmental changes, which were simulated by quantitative and qualitative manipulation of element inputs. The effects of an improved immission quality which can be expected as a result of implementation of air protection measures, were investigated in a deacidification experiment. The effects of long periods of drought phases were tested in a drying out experiment.

Internationally, the experiments were integrated into the framework of the projects EXMAN (Experimental Manipulation of Forest Ecosystems in Europe, project duration 1987-1995, Bredemeier 1995, Wright and Rasmussen 1998) and NITREX (Nitrogen Saturation Experiments) supported by the EU. In this research co-operation similar projects were carried out on the Danish west coast (Klosterheede), in south-western Ireland (Ballyhooly), in the Netherlands and in Höglwald in Bavaria.

The project was co-ordinated by the Forest Ecosystems Research Center of Göttingen University and the work carried out by groups in the Institute of Soil Science and Forest Nutrition, the Institute of Silviculture and the Zoological Institute. The results presented here focus on the work of the Ecophysiology and Growth group, which investigated aboveground reactions of the trees to the manipulations.

2.1.2 Materials

2.1.2.1   *Investigation area and experimental site*

The experimental sites of the roof project are in the department 4257j of the forestry administration Dassel, Lower Saxony (former forestry administration Neuhaus dept. 257j), about 50 km north-west of Göttingen, 510 m above sea level (51° 46' 09" N, 9° 34' 52" E). The suboceanic climate prevalent in the area *Hoher Solling* is characterised as cool humid. The average temperature of the vegetation period (May-September) is 13.5 °C, almost twice as high as the annual mean (Fig. 2.1-1). About 120 days were with frost (temperature minimum below 0 °C). The relatively high amount of precipitation is evenly distributed over the course of the year. December the month with the highest precipitation of 105 mm exceeds February the month with the lowest rainfall by only 30 mm. Long term measurements showed marked differences between the years: The annual sums fluctuated over the past 30 years between 400 mm (1959, 1983) and 1500 mm (1960, 1970) (Gravenhorst and Szarejko 1990). Deviations from these long term mean values during the investigation over 7 years were up to 25% for annual precipitation (1994) and about 1 °C for the annual mean temperature (1990, 1992, 1994, 1996) (Fig. 2.1-1).

For the years with intensive experimental and measurement programmes the following specific conditions were relevant: In 1993 the course of the temperature curve was even with two periods of light frost in February and November, and a very warm spring starting already in April. The summer temperature curve was temperate without major peaks. The precipitation was also evenly distributed over the year. Only in July (153 mm) and December (204 mm) markedly more rain fell. In 1994 temperatures only increased slowly during the spring months after a sudden decrease in February, as from March to April the rainfall of 320 mm was above average. July with 19.9 °C was the warmest month of the total 4 year measuring period. The temperature of the vegetation period was also about 1.6 °C higher than the annual mean. Despite of the very dry vegetation period (406 mm) the highest amount of annual rainfall over the time period of the investigations was measured with 1304 mm. The year 1995 with a mild February and high precipitation was a warm year. With 18.5 °C it was not as hot as in 1994, but August was the warmest month of the measuring period with 16.8 °C. With 507 mm rainfall, only 44% of the total precipitation occurred during the vegetation period. In general, 1995 was drier than the previous year as October with 10.9 °C, was very warm and dry with only 24 mm precipitation.

In 1996 a very cold frost period from January to March was followed by a dry early summer. Due to the rainfall in July and particularly in August 44% of the precipitation occurred during the vegetation period. In general the year was drier than normal, and with a mean temperature of 12.3 °C over the vegetation period it was colder than normal.

Measurements of air pollutants showed that the SO_2-pollution was high during winter months. It reached an average concentration of more than 0.1 mg m^{-3} which is comparable to the conditions in densely populated regions. Ozone was determined in high concentrations of more than 0.1 mg m^{-3}. The average nitrogen concentration in the air during the winter months was mostly more than 0.05 mg m^3. In total the sulphur input has considerably decreased (Fig. 2.1-2). After a maximum input was reached in the middle of the 70s with more than 100 kg ha^{-1} y^{-1} it decreased to below 50 kg at the beginning of the 90s and today to just above 30 kg. In contrast, the total amount of nitrogen deposition, composed almost of equal amounts of ammonium and nitrate nitrogen, increased over the same period of time from just 30 kg ha^{-1} y^{-1} to 40 kg.

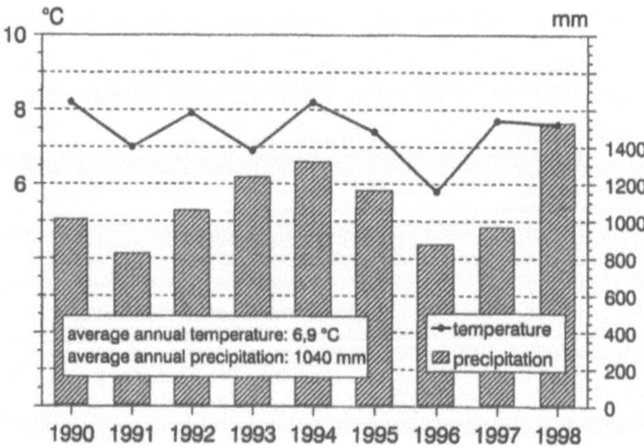

Fig. 2.1-1. Average temperature and precipitation development during the observation period.

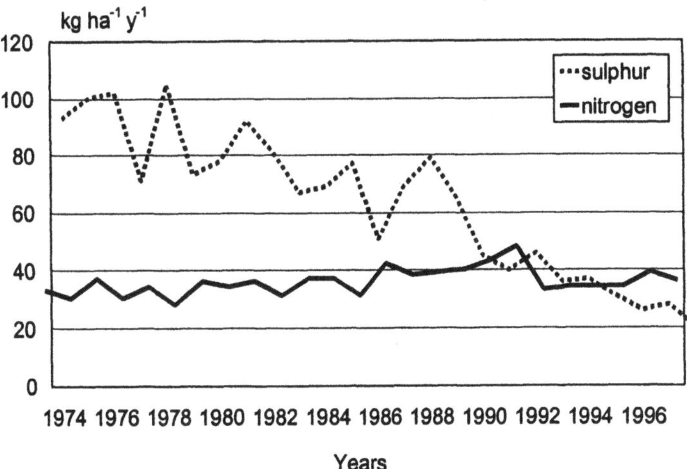

Fig. 2.1-2. Sulphur and nitrogen input in the throughfall in the time period 1974-1996.

The experimental sites are on the slightly sloped Solling Plateau. The geological parental material is a Triassic red sandstone on which slightly podsolic, weakly pseudogleyic brown-earth layers have formed (Heupel 1989). The nutrient potential of the sites is mainly determined by loess layers of a varying thickness. In the investigation area the loess is up to one meter deep, but shows large differences over small spatial areas (Blanck et al. 1993). The organic layer varying in thickness between 6 and 9 cm, corresponds to an average dry substance of 114 t ha^{-1} (Heupel 1989), of which just over half of the total amount can be allocated to the O_L an O_F layer. Probably due to the high atmospheric nitrogen inputs, the C/N-ratio of 25 found in all humus layers is less than that normal for the fine-humus-rich moder humus form. The low magnesium and calcium contents in the humus layer are evidence of the generally poor nutrient conditions (Table 2.1-1).

Table 2.1-1. Average storage of nutrients in the humus and mineral soil up to a depth of 80 cm (Data from Lamersdorf 1998).

Element storage	C	N	P	K	Ca	Mg	Mn	Fe	Al
in humus (t ha^{-1})	48	1.9	0.11	0.22	0.16	0.08	0.02	0.91	1.0
in the mineral soil (t ha^{-1})	55	3.5	0.86	1.14	0.46	0.17	0.98	0.85	13.7
sum (t ha^{-1})	103	5.4	0.97	1.38	0.62	0.25	1.00	1.76	14.7
proportion in the humus (%)	46	35	11	16	26	31	2	52	7

The very low pH-values in the upper soil of around 3 (pH $CaCl_2$) are within the aluminium, and iron buffer ranges (Ulrich 1983a). The pH increases to values of more than 4 at deeper soil depths. As a result the contents of sodium, potassium and magnesium in the mineral soil at all soil depths are very low, contributing only 6% to the total cation exchange capacity. The highest amounts are found in the soil layers at 20 to 40 cm depths. Relatively high amounts of some nutrients have accumulated in the organic layer: nitrogen and magnesium contribute one third and calcium a quarter to the total amount.

2.1.2.2 *The spruce stand*

For the experiment a 57-year-old spruce (*Picea abies* Karst.) stand was selected, at only a few hundred meters distance from the experimental sites B1 and F1 used for the long term investigations of spruce and beech which started in the 1960s (Ellenberg et al. 1986). These investigations provided information about the pollutant input to the sites through long-term measurements. Table 2.1-2 shows the present average element inputs.

Table 2.1-2. Average annual element inputs (kg ha^{-1}) in the stand via precipitation at the control site D0 (mean of the time period 1990-1994).

Na	K	Ca	Mg	Fe	Mn	Al	H	NH$_4$-N	NO$_3$-N	SO$_4$-S	PO$_4$-P	Cl
18.7	26.1	17.4	3,9	0.4	3.0	1.0	1.1	17.6	18.9	42.4	0.2	36.5

The spruce stand is the second generation of this tree species, which replaced the natural wood-rush/beech forest *(Luzulo-Fagetum)*. The spruce stand was planted in 1933 and as a result of several silvicultural measures was thinned to 900 trees ha^{-1} by the beginning of the project (1990). The stand was then 57 years old and had an average DBH of 27 cm where the strongest trees already exceeded 40 cm. The mean height of the stand was 19.7 m in which the highest tree measured 25 m. The h/d ratio, the quotient calculated from tree height and DBH used to determine the stand stability, shows a favourable average value of 73. The average annual increment of 9 m^3 ha^{-1} corresponds to a yield class of II.2. Almost all trees showed old peeling scars at the stems caused by red-deer, noticeable to varying degrees as wound occlusions. At the start of the experiments the spruce were allocated to the damage classes 2 (medium damage) and partly damage class 3 (severe damage). In addition to needle loss, older needles were chlorotic.

2.1.3 Methods

2.1.3.1   Experimental design

The spruce stand was divided into several experimental sites, of which the three sites D1, D2 and D3 were roofed in order to be able to manipulate the water and element inputs. The roofs are self-supporting wooden structures spanning over 17 m and with a 3.5 m ridge height. A central maintenance building was built on concrete foundations. Each roof is covered with transparent polycarbonate sheeting and covers a ground area of 300 m². The total precipitation falling on the roofs in the stand is directed by pipes to collecting tanks in the maintenance building. Here the chemical composition of the water can be manipulated by an installed desalination and subsequent dispensing equipment. It is also possible to deviate water for a temporary storage in the storage tanks (42 m³ ≈ 140 mm precipitation). Finally the precipitation of the stand – depending on the roof area and experiment in natural or chemically changed form – is transported via a pipe system back underneath the roofs and released as rain using sprinklers. The three quadrangular roof structures could be used from spring 1991.

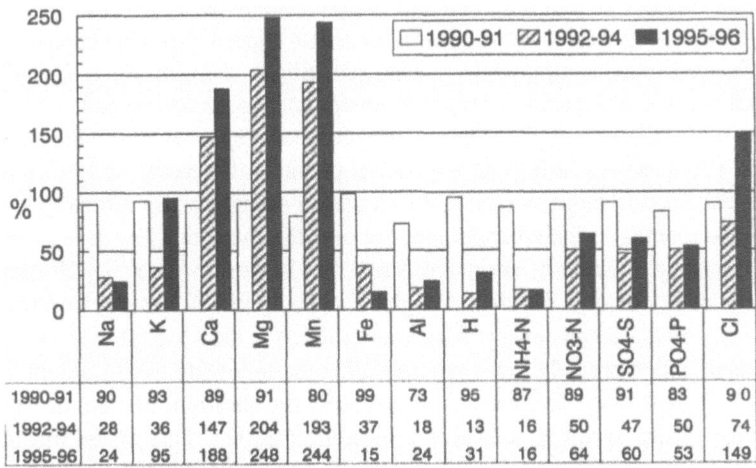

	Na	K	Ca	Mg	Mn	Fe	Al	H	NH4-N	NO3-N	SO4-S	PO4-P	Cl
1990-91	90	93	89	91	80	99	73	95	87	89	91	83	9 0
1992-94	28	36	147	204	193	37	18	13	16	50	47	50	74
1995-96	24	95	188	248	244	15	24	31	16	64	60	53	149

Fig. 2.1-3. Relative element concentrations (%) under the de-acidification roof D1 in comparison to the control site D0.

2.1.3.2 *Experimental treatments*

Under the de-acidification roof (D1) an unchanged amount of precipitation but in a changed form was sprinkled. In doing so the conditions were to be simulated which compared to the composition of pre-industrial precipitation. In order to attain this result the water was first de-mineralised in the desalination device and subsequently a nutrient solution and sodium hydroxide was added, thus the manipulated crown edge only contained half of the normal concentrations of sulphate, nitrate and phosphate (Fig. 2.1-3). Considering the severely reduced ammonium nitrogen content to 16% of the normal concentration, this manipulation reduced the total nitrogen input to almost one third. At the same time the pH-value of the throughfall was increased from an ambient 4.1 to between 6.0 and 6.4, while the contents of protons, aluminium and iron ions were markedly reduced (Fig. 2.1-3). The similarly strong increase of the calcium and magnesium inputs certainly does not correspond to a simulation of pre-industrial inputs. However, it means an optimisation of the site conditions as it can be expected to result from a reduction of the pollutant inputs in connection with soil amelioration measures (liming, fertilisation). Figure 2.1-3 shows the relative change compared to the control for three temporally subsequent phases. The period 1990/91 is the preliminary phase during which almost the same concentrations registered for the control area were also dispersed under the de-acidification roof. The second period is the intensive measuring phase, during which the investigations of the trees and other parts of the ecosystems were carried out at high temporal and spatial resolutions. In general the final phase 1995/96, during which measurements were also carried out, was supposed to continue the previous manipulations unchanged. Slight changes are technically induced and not to avoid, greater changes represent intentional corrections.

Another roof (D3) was used for the investigation of responses to drought (Fig. 2.1-4). The precipitation during the vegetation period in the years of 1991 until 1994 were collected in large storage tanks an after a drought phase normally lasting for several months sprinkled under the roof over the space of a few days (Table 2.1-3). The average amount of sprinkled water was 10 dm^3 m^{-2} day^{-1} (= 10 mm), however, the daily amounts differed strongly. Especially in 1992 strong variations occurred: Very high amounts of sprinkled water such as at the 11. Sept with 28 mm were corrected with extremely low amounts during the following day (1 mm at the 12. Sept.). This experiment was carried out to clarify the question, whether drought and rewetting phases result in intensive acidification pushes. Due to the marked drought stress responses observed at the trees in the years of 1993 and 1994 subsequent to a drought over several months in 1995 no further drought experiments were carried to give the stand a chance to recover, instead the phase of recovery was monitored by continuous measurements.

Table 2.1-3. Artificial drought phases and rewetting amounts under roof 3.

	1991	1992	1993	1994
drought	02.09.-13.10.	15.05.-06.09.	01.04.-19.09.	01.04.-17.07
duration (days)	42 d	108 d	172 d	110 d
sprinkling	14.10.-25.10	07.09.-16.09	20.09.-08.10	18.07-19.08
duration (days)	9 d	12 d	17 d	24 d
amount of sprinkled water	$79\ l\,m^{-2}$	$108\ l\,m^{-2}$	$193\ l\,m^{-2}$	$184\ l\,m^{-2}$
amount/day	$8.8\ l\,m^{-2}$	$12.0\ l\,m^{-2}$	$8,2\ l\,m^{-2}$	$7.7\ l\,m^{-2}$

A directly adjacent non-roofed part of the stand (D0) and the so called zero roof (D2) served as the controls. Here the collected precipitation was sprinkled without changing the amount or the composition and thus the environmental conditions imulated. Thus it was possible to test the validity of probable "roof effects".

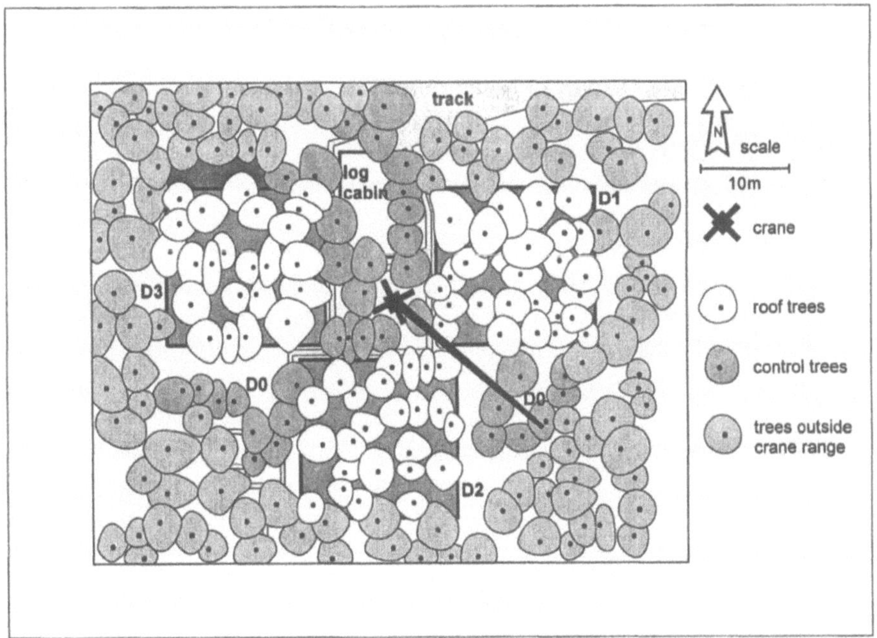

Fig. 2.1-4. Ground plan of the roof project.

In the centre of each of these four experimental variants within a measuring unit of 9 m^2 lysimeters were installed at a soil depth between 10 and 70 cm which since 1989 ensured a continuous supply of the soil solutions and their analysis. Using tensiometers and since 1993 additionally installed TDR-probes it was possible to collect continuous data about the soil tensions (Blanck et al. 1993). The precipitation rates of the crown edge have been registered for the non-roofed control site D0 by an automatic rain monitor device at quarterly hour intervals since November 1992. For the chemical analysis of the crown edge for the first time in 1989 and then regularly from 1994 snow buckets were installed at the control site. In order to ensure a continuous investigation of the fine root development keeping damage at a minimum, special procedures were devised, using an endoscope to monitor several meters of pipe systems made of transparent perspex pipes at different soil depths (Murach et al. 1993). Inventories of the development of the ground fauna were carried out regulary by by the Zoological Institute of the University Göttingen (Alphei et al. 1993).

An important objective was the observation of responses by the trees to changed environmental conditions. In order to carry out the necessary measurements in the crown area, in spring 1992 a crane 30 m high was installed in the center of the roofed area. This was equipped with a special transport system for persons (a cabin with a floor space of 100 x 70 cm) which made it possible to reach the crown area of all of about 100 trees which belonged to the experimental sites. With this method regular physiological measurements of the carbon budget (photosynthesis, respiration) and the water budget (transpiration, xylem water potential, osmotic potential) as well as the shoot elongation were carried out. In addition each year needle samples were taken for element analyses and also an assessment of the condition by visual observation of the needle losses and needle chlorosis. The radial growth of the trees was determined at monthly interval by permanently fixed circumference measuring bands around the trees. Selected sample tree were additionally equipped with a home made microdendrometer which registers changes in radial growth of the stem at a differentiated level providing information about the growth and water budget of the trees.

The roof project was started after a two year planning phase at the end of 1989. Until 1991 the within a preliminary phase input data and soil solution data were regularly taken without any experimental impacts. The installation of the roofs took place in summer 1991 and one year later the crane was in use for the investigations of the crown area.

Acknowledgement

The study received financial support from the German Federal Ministry of Research and Technology (BMBF, 325-7291 OEF 2019) and from the State of Lower Saxony.

References

Alphei J, Bonkowski M, Koch M, Schauermann J (1993) Reaktionen von Mikroflora und Tieren des Bodens auf experimentelle Manipulationen des Niederschlagswassers eines Fichtenforstes. Forstarchiv 64:194-201

Blanck K, Lamersdorf N, Bredemeier M (1993) Bodenchemie und Stoffhaushalt auf den Dachflächen im Solling. Forstarchiv 64:164-172

Bredemeier M (1995) Experimental manipulations of the water and nutrient cycles in forest ecosystems (patch-scale roof experiments: EXMAN). Agr. Forest Meteorol. 73:307-320

Bredemeier M, Blanck K, Dohrenbusch A, Lamersdorf N, Meyer AC, Murach D, Parth A, Xu YJ (1998) The Solling roof project - site characteristics, experiments and results. Forest Ecology and Management 101:281-293

Bredemeier M, Lamersdorf N, Murach D, Dohrenbusch A, Alphei J (1999) Das Dach-Projekt Solling - Gesamtwertung des Ergebnisstandes. AFZ/Der Wald 54:70-71

Dohrenbusch A (1996) Das Dachprojekt - Ein Versuch, die Auswirkungen und Wirkungsmechanismen von Umweltveränderungen auf Waldökosysteme zu verstehen. Tagungsbericht der Jahrestagung des Deutschen Verbandes Forstlicher Forschungsanstalten/Sektion Waldbau, Schopfheim-Wiechs, 17.-19. September 1996:21-29

Dohrenbusch A, Jaehne S, Meyer AC (1999) Reaktionen eines Fichtenaltbestandes auf ein verändertes Wasser- und Nährstoffangebot. AFZ/Der Wald 54:60-62

Ellenberg H, Mayer R, Schauermann J (1986) Ökosystemforschung – Ergebnisse des Solling-Projekts. Ulmer, Stuttgart

Gravenhorst G, Szarejko Z (1990) Das Klima des Sollings (in: Exkursionsführer Solling - Oktober 1990). Ber Forschungszentrum Waldökosysteme, Göttingen, Reihe B, Bd 17:5-14

Heupel GM (1989) Bodenkartierung der Sollingversuchsflächen Abt. 257. Unveröff. Manuskript, Inst. f. Bodenkunde u. Waldernährung, Göttingen

Lamersdorf N (1998) Auswirkungen wiederholter Bodenaustrocknungen auf den Stoffhaushalt eines Fichtenwald-Ökosystems im Solling. Habilitationsschrift, Inst. f. Bodenkunde u. Waldernährung d. Universität Göttingen, Göttingen

Murach D, Ilse L, Klaproth F, Strunk P, Wiedemann H (1993) Zum Einsatz verschiedener Wurzelbeobachtungsmethoden im Dach-Experiment im Solling. Forstarchiv 64:185-188

Ulrich B (1983) Stabilität von Waldökosystemen unter dem Einfluß des „sauren Regens". Allg Forstz 38:670-677

Wright RF, Rasmussen L (1998) Introduction to the NITREX and EXMAN projects. Forest Ecol. Manage., NITREX/ EXMAN special issue 101:1-7

2.2 Effects of nutrient and water supply on growth and seed production of a Norway spruce stand

A. Dohrenbusch, S. Jaehne, A.C. Meyer

Institute of Silviculture, Göttingen University, Büsgenweg 1, D-37077 Göttingen, Germany, e-mail: adohren@gwdg.de

Abstract

During a ten years lasting period the development of increment of a 60-year old spruce stand was examined. The environmental conditions of this stand were manipulated significantly. After long terms of drought distinct reactions of the trees were visible in growth and nutrition. The reactions of height-increment were more distinctly than the effects on diameter-increment. Furthermore, the trees of the dominating social classes (Kraft I and II) reacted more on low water-supply than the dominated trees. So it is probable that a long lasting stress by drought effects the stand structure, too: the vertical structure of a stand gets more homogenous and the diversity in the stand structure decreased. The effects of a reduced input of sulphur and nitrogen resulted in a more favourable composition of nutrients in the needles.

Key words: roof project, *Picea abies*, de-acidification, acid rain, clean rain, drought, height growth, diameter increment, fructification

2.2.1 Introduction

Growth and fructification of trees are characteristics, which can usually be quantified and assessed with simple methods. Apart from these methodological advantages, these are also the criteria of particular interest for applied forestry. This considered, it seemed logical to investigate the influence of changed environmental conditions on the development of growth and on the regeneration ability. For the drought experiments only the tendency of the expected reactions was predictable, but not their extent. In contrast, the prognosis for yield in the de-acidification experiment was difficult as it was not clear, whether the positive effects of a decreased input of pollutants would be compensated by a strong reduction of nitrogen.

2.2.2 Material and Methods

Over a total period of nine years, from all 74 spruce trees of the three roofed sites yield data were collected (27 trees from roof 1, 24 from roof 2 and 23 from roof 3). The control tree group analysed from the start of the project, but did not grow within the range of the crane, were thus replaced by a new control group in 1995. These 35 reference trees (16 original control trees at the side of the roofs as well as 13 trees close to the house and 6 close to the crane foundation) are all within reach of the crane. The radial growth was measured with radial measuring bands, which were permanently fixed at a tree height of 1.3 m in 1989. Using this method the radial growth of each tree was registered monthly with an accuracy of ± 0.2 mm. The annual radial increment was estimated using the data obtained in October. By October the transpiration rate of the trees was already markedly reduced, so that expansion and shrinking processes of the stems did not play an important role.

The measurement of the annual height increment was also carried out in October. In order to do this the lengths of the newly grown apical shoots had to be determined. This was not possible until the crane could be used in 1992. For the previous years after 1988 the height increment could be estimated on the basis of the distance between the branching nodes. The total height of the trees was first measured in October 1992 at the beginning of the experiments. The subsequent annual height increment rates were used to update this measurement.

The reproduction rate of the trees was determined by counting the cones in autumn. The crane was used for this work, and also for the visual assessment of the crowns at different heights and perspectives. An overview of the data are shown in Table 2.2-1.

Table 2.2-1. Investigations carried out.

Parameter	Start	Interval	Number
Radial growth	1990 roofs / 1995 control	monthly	109 trees
Height growth	1988	annually	109 trees
Fructification	1992	annually	109 trees

2.2.3 Results

2.2.3.1   Height growth

The course of the annual height growth showed similar trends for all experimental variants. The mean height increment decreased continuously since the beginning of the measurements in 1988 from an average of 37 cm on all sites to a minimum in the fifth year of monitoring in 1992 (Fig. 2.2-1). At this date the mean shoot length was only 14 cm. After which, up to 1996, a marked increase in growth was again observed.

In the years 1993 and 1994, an influence of the experimental treatments on the height increment of the trees was shown. As a result of the long dry periods during the summer months of previous years the mean height increment on the D3-site was significantly reduced by about half compared to the other sites. After 1995 on the basis of all trees, no effects were shown induced by the drought in previous years. By contrast, the effect of the de-acidified precipitation on the trees of the D1 site for the total period monitored were statistically not significant.

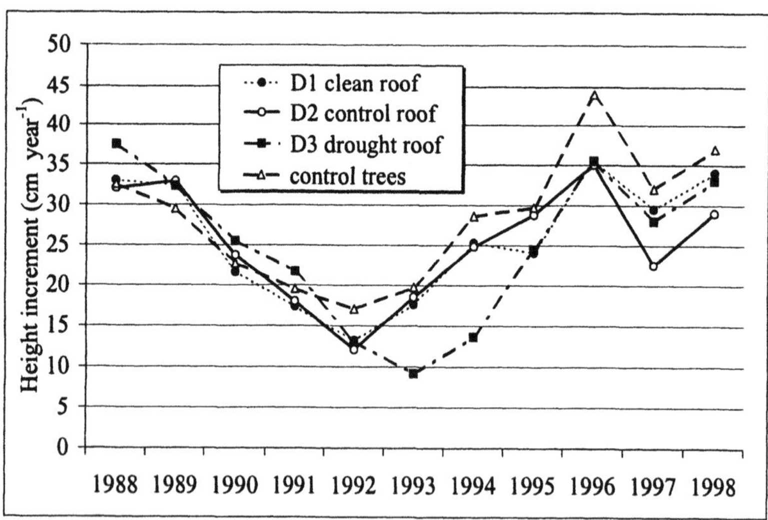

Figure 2.2-1. Annual height growth.

2.2.3.2 *Radial growth*

Figure 2.2.-2 clearly shows that the pattern of radial growth at a height of 1.3 m does not correspond to the course of the height increment. During the whole monitoring period of nine years not even a trend towards a change is detectable. The highest radial growth increment was determined in 1997 for most of the sites. In general, the non-roofed control trees showed a higher radial growth than the roofed trees. That the control trees are first shown in 1995 is as the control trees used until then belonged to a different collective.

It was not possible to show conclusively an effect of the treatments on the radial growth of the trees. Although the drought experiment differed from the control under the roof during the years of intensive drought by an average of more than 0.5 mm radial increment, this was not statistically significant at any point in time. Only the differences determined for the year 1995 between the de-acidification and the drought experiment were significant.

Marked regeneration effects were observed in the annual radial growth of the trees under drought conditions. During 1993 to 1995 the radial growth compared to the control was still significantly reduced, while in 1996 the growth of the trees exposed to drought was markedly lower with 1.9 mm compared to an average of 2.6 mm of other sites. In 1996 the growth on the de-acidification site with 2.8 mm was highest, the difference at all sites was not statistically significant. From the cumulated monthly increment rates since 1989, it is noticeable that droughted trees show

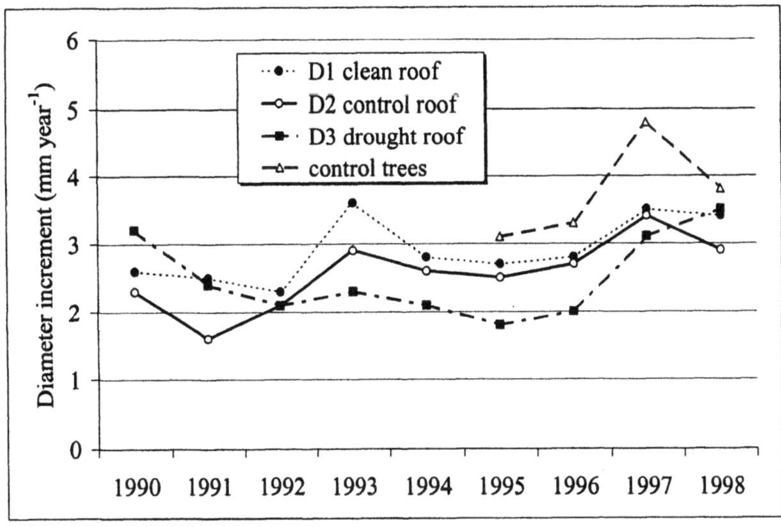

Figure 2.2-2. Annual diameter increment.

a strongly reduced growth after 1993. The trees of the de-acidification experiment showed an increased radial growth of the stem. However, this improvement was not statistically significant at any point in time.

2.2.3.3   Effects on stand structure

Based on the differentiation of the data material according to tree classes, an effect of the experimental treatments on tree growth could be shown. This was based on the hypothesis that non-dominating trees are less affected by changes in the abiotic site factors. On the one hand they are exposed to smaller amounts of immission than the larger trees. On the other hand the competitive conditions probably represent a stronger limiting factor for their growth potential. Thus a worsening of the environmental conditions (compare D 3) or an improvement (compare D1) will have lesser effects than for dominating trees.

Figures 2.2-3 and 2.2-5 show the mean values for height and radial increment of the dominating and co-dominating trees (tree classes 1 and 2 based on Kraft). The values for the trees, which are at least partially overshadowed (tree classes 3 to 5), are shown in Figures 2.2-4 and 2.2-6. The classification of the trees according to their sociological order was based on the stand condition, before the two experiments began in 1990.

The effects of the drought experiment were clearly observable in both the dominating and co-dominating trees from 1993. At the D3-site the lack of growth was significantly compared to all other sites. On average the height increments (Fig. 2.2-3) of the trees were reduced by 50% and the increments of radial growth (Fig. 2.2-5) by 20%.

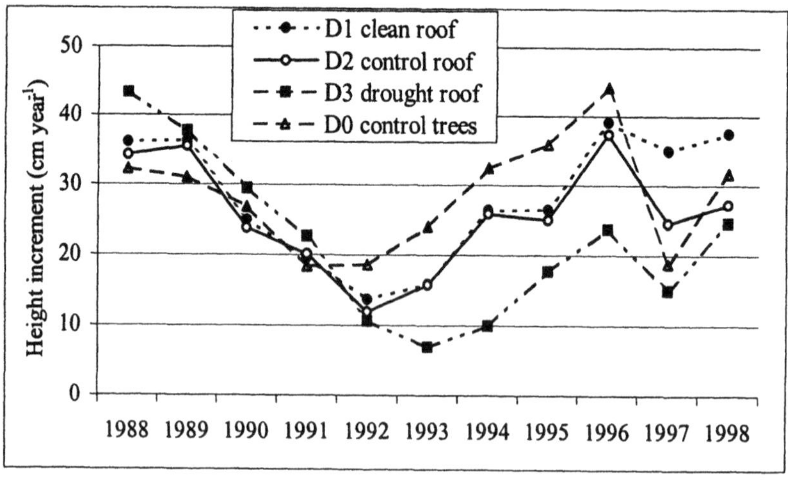

Fig. 2.2-3. Height increment for the dominating trees (tree classes 1 and 2 based on Kraft).

This development could be observed over a period of four years up to and including 1996. Thus the influence of the drought experiment, which was finished in 1994 on the D3-site, continued to be effective for two years after the treatment was stopped. The trees showed no signs of a quick regeneration. A visual assessment of the tree crowns also suggests that although a regeneration process had actually taken place, the damage in some cases is however irreversible.

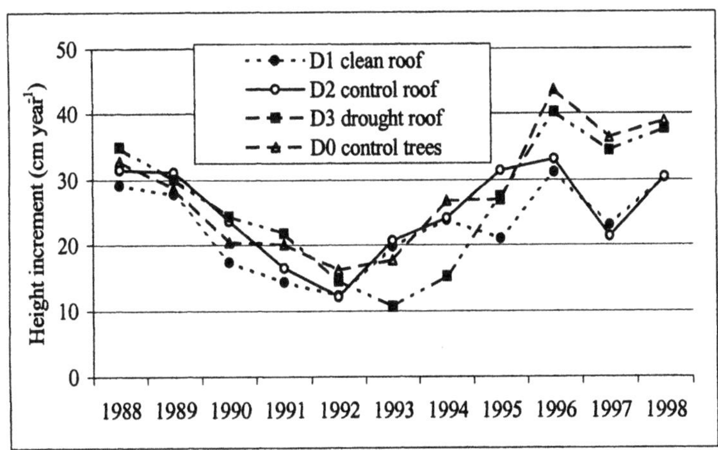

Fig. 2.2-4. Height increment for the dominated trees (tree classes 3 to 5 based on Kraft).

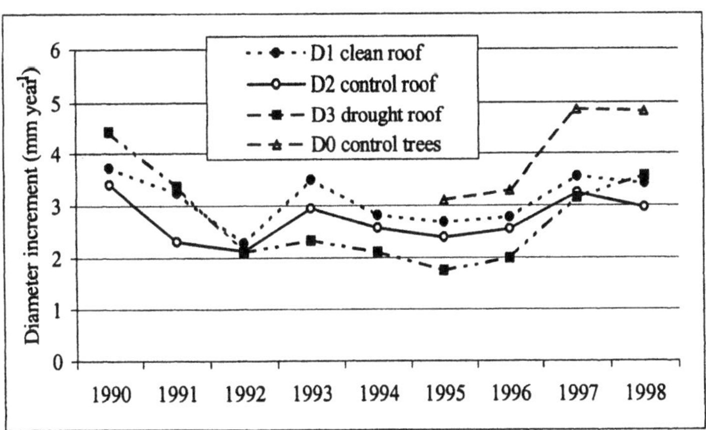

Fig. 2.2-5. Diameter increment for the dominate trees (tree classes 1 and 2 based on Kraft).

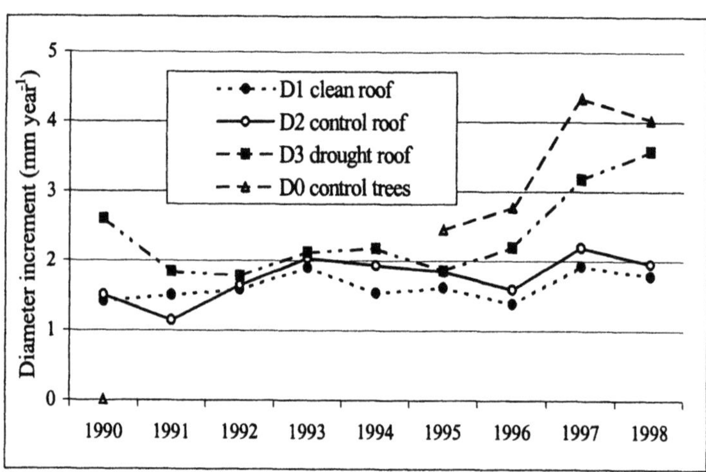

Fig. 2.2-6. Diameter increment for the dominated trees (tree classes 3 to 5 based on Kraft).

For the dominated trees on the D3-site a reduced increment in height development was only determined (Fig. 2.2-4) during the years of severe drought (1993 and 1994). However, in the absolute volumes there were marked differences, which were statistically not significant. After a fast adjustment of the shoot length growth to the values of the control trees (Fig. 2.2-4) as early as one year after the drought, the dominated roofed trees on D3 developed better in the following years than the trees of the other two roofs. The radial growth of the dominated trees was not affected by the experimental treatment (Fig. 2.2-6).

The results obtained have confirmed the hypothesis, that a worsening of the environmental conditions mainly affects the prevalent and dominating trees of a stand, while the dominated trees hardly show any reaction. If the conditions continue over several years the structure of the stand may become more homogenous. However, a complete adjustment and formation of stands composed of one growth layer is unlikely to occur.

The evaluation of the results obtained from the de-acidifying experiment did not show any significant effects on tree growth. The development of the height and radial growth increments on the D1-site was similar to the development on the control site D2. The investigation based on the sociological classes could also not show any differences between the dominating and dominated trees. The absence of a reaction is probably due to the compensating effects of the changes in the input. Although as a result of the de-acidification a better nutrient supply and thus a higher growth rate were expected, the high reduction in nitrogen inputs may have had the opposite effect. In addition on the D1-site, which has a total of 27 trees, providing a much smaller rooting area for each tree, which may have resulted in a stronger competition than on the other roofed sites (with 24 or 23 trees respectively per 300 m²). Also comparing the density dependent basal area, the value for D1 with 56 m² ha⁻¹ is 10% higher than that of the two other sites. These comparatively very high stocking rates are also the result of bark stripping damage, which occurred on almost all stems. Wound occlusion leads also to the formation of asymmetrical stem cross sections and hollows, which prevented a more accurate determination of the diameter of the stem. The typical symptoms of thickened trunk bases due to butt rot damage from *Heterobasidion annosum* induced by bark stripping damage also lead to systematically increased DBH.

2.2.3.4   *Saisonal growth development*

The permanently fixed rings for radial growth measurements permitted not only the determination of the annual values but also the monthly changes. In the years of 1993 and 1994 on D3 for several months of the vegetation period strong drought conditions were simulated. Figure 2.2-7 shows the mean growth values of the radial growth.

In both years by far the highest growth rate was determined in June, July and August. In 1994, a relatively wet (20% more precipitation above the long term average), and very warm year (+1 °C above the long term average), changes in radial growth were shown from May onwards. the control trees outside the roofed area had a higher annual growth over the whole year which began at the start of the vegetation period

The effects of the extreme drought in 1993 lasting from April to September did not become apparent until July, when the radial growth was markedly reduced. Compared to the other roofed trees the increment in July was lower by one third. In August growth stagnated to 0.2 mm, while the values at the other sites varied between 0.6 and 0.9 mm.

Towards the end of the vegetation period in September, a seasonally related decrease in growth of all trees decreased to the low levels of the droughted trees was shown. However, after an intensive rewetting was carried out an opposite reaction began. It may be assumed that the trees on the D3-site in September still had an unused growth potential which had not been activated during the drought. This could be used to compensate for a part of the losses in growth. To what extent the increase in radial growth by 0.4 mm in October 1993 was related to an actual gain in growth or only to temporary swelling processes of the stem and bark could not be determined. When the drought experiment was repeated in the following year, it was terminated by rewetting in July 1994. However, although climatic conditions were very warm, no extreme changes in growth were determined for the trees on the D3 site at the peak of the summer. Rather, a continuous reduction of growth was observed lasting over the whole of the vegetation period.

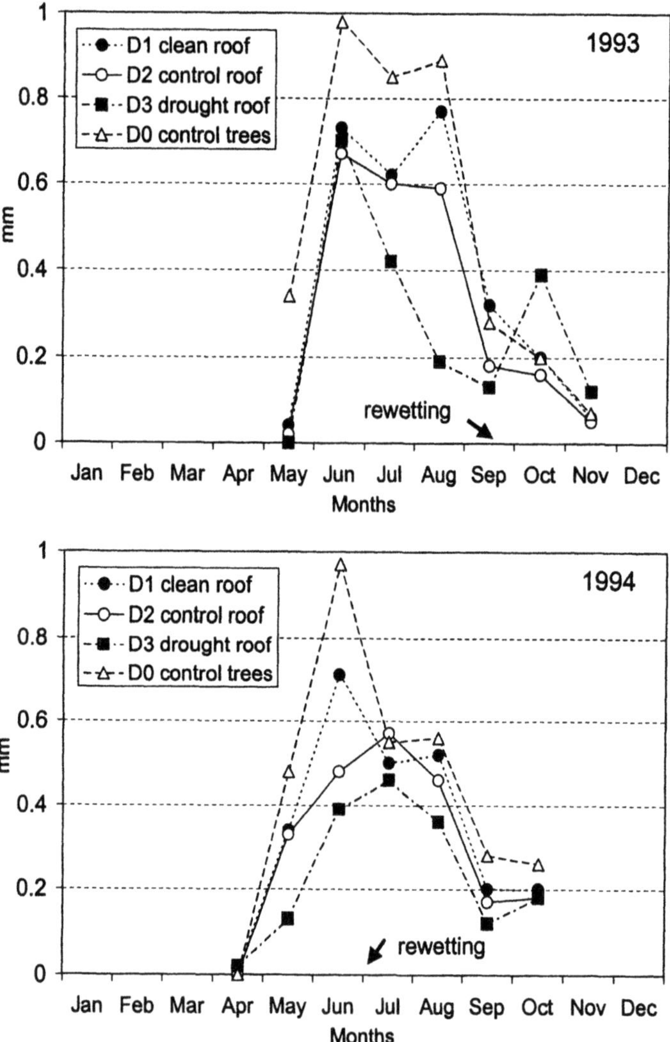

Fig. 2.2-7. Monthly diameter increment in 1993 and 1994.

2.2.3.5   *Fructification*

Figure 2.2-8 shows the average number of cones in the years 1992 to 1998. The first count in late summer 1992 showed a large number of cones with an average between 93 and 97 per tree. However, the mean values on an area basis conceals marked differences between individual trees. While some trees had several hundred cones, others had none. During the following two years the amounts dropped to 30 cones in 1993 or 5 cones 1994, respectively. After a new increase in the years 1995 and 1996 the number of cones in 1997 reached similar amounts to those of 1994.

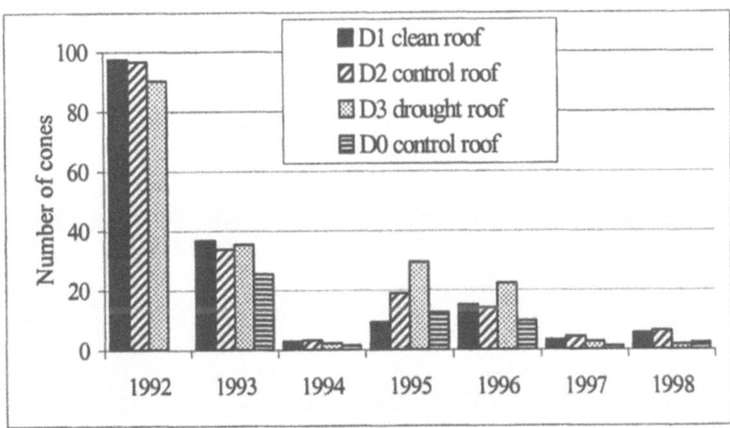

Fig. 2.2-8. Number of cones in the years 1992 to 1998.

During the total monitoring period no significant differences were found between the experimental sites. Independent of the site however, there is a close relationship to the sociological order of the tree, and thus the parameter DBH. A highly significant, negative correlation was determined between the annual height growth and the fructification intensity in the same year. The higher the number of cones formed during the vegetation period, the smaller was the growth in height of the trees.

2.2.4 Discussion

The growth increment of trees must be considered to be the result of several factors which underlie complex interrelations. Thus it is very difficult to investigate individual aspects and their effects separately. This applies especially to the investigations of environmental changes carried out in the roof experiments. A result is that continuous drought stress resulted in marked increment losses. In contrast, the amelioration treatments of the soil chemical conditions carried out in the de-acidification experiments resulted in an increment increase especially in dominant trees. It has often been observed that water supply is a stronger influence than nutrient supply if site conditions are improved (Nilsson and Huttl 1977).

The effects of drought stress were investigated by Wiedemann (1925) in several medium aged spruce stands in Saxony. A relationship was determined between the observed increment losses in the trees and the number of months with drought (precipitation of less than 40 mm) during the vegetation period. This corresponds with the results of Gross (1988) who determined a significantly reduced increment growth rate under drought stress conditions in 10- to 15-year-old spruce trees. A decrease in shoot length growth in 4- to 5-year-old spruce trees after a drought period was shown by Michael et al. (1989). Nilsson and Wiklund (1992) describe a reduction of needle size as a direct result of drought stress. Gross and Pham-Nguyen (1987) relate this process to the shorter shoot lengths. In addition, effects ranging from a thinning of needles to a total loss of older needle generations may occur. Thus it may be concluded that the rate of photosynthesis decreases in spruce trees exposed to drought stress. This is postulated by Gross (1988) and Gross and Pham-Nguyen (1987). However, not only the inhibition of the assimilating system has a negative effect on the increment rate of the trees. It must also be assumed that the growth of the root system is reduced or altered (Blank et al. 1995). As particularly the fine root system is affected, water and nutrient uptake by the trees is decreased. On the D3-site the radial and height increment regenerated within two years subsequent to the termination of the drought experiment. Considering the severe damage in some trees the regeneration time seems remarkably short. Wiedemann (1925) investigated spruce stands and reports a time span of 2 to 20 years before a regeneration of the increment rate sets in.

A comparison of the de-acidification site D1 with the roofed control site D2 allows conclusions to be drawn about the effects of soil acidification. Despite a markedly stronger intraspecific competition (higher density of the stand at the beginning of the experiment) the trees on the de-acidification site showed continuously better growth. This is most certainly due to the experimental treatments carried out which improved the nutrient supply to the trees. Widstrom and Ericsson (1995) emphasise the importance of nitrogen and magnesium for the growth of spruce and birch. Both elements play a key role under the prevailing site conditions in the higher Solling uplands (Mohren et al. 1993). The consequences resulting from magnesium deficiency are reported in Mehne-Jakobs (1995). Here it is assumed that as a result of the reduced transport of assimilates in the trees growth is inhibited. It also appears that the formation of chlorophyll strongly depends on the magnesium supply to the needles.

An assessment of the importance of nitrogen for the growth rate of spruce trees is more problematic. On the one hand, as Rosengren-Brinck and Nihlgård (1995) point out, an increase of nitrogen input provides better growth conditions, but at the same time it might represent a stress factor for the trees. However, some site and regional differences render it difficult to determine the amount of nitrogen available (Evers and Moosmayer 1980). An increase of the nitrogen supply thus does not automatically increase the increment rate. On the D1 site the increment rate even increased although the nitrogen inputs were shown to be markedly reduced. Only on the sites with an insufficient supply of nitrogen can specific fertilisation treatments with nitrogen result in an increase of the growth rate. This was shown by Nilsson and Wiklund (1992) in a 25 year old spruce stand in southern Sweden. For sites with a sufficient N-supply a balanced level of nutrient elements is required, independent to a large extent of the total amount of available nitrogen (Nilsson et al. 1993). This seems to be confirmed by the results obtained in the de-acidification experiment on the D1-site.

Acknowledgements

The study received financial support from the German Federal Ministry of Research and Technology (BMBF, 325-7291 OEF 2019) and from the State of Lower Saxony. We want to give our special thanks to the technicians G. Brauer, A. Nannen, K.-H. Obal, D. Pryor, U. Schmidt, C. Suner and H. Zimmer for their collaborations and O. Godbold for improving the English.

References

Blanck K, Lamersdorf N, Dohrenbusch A, Murach D (1995) Response of a Norway spruce forest ecosystem to drought/rewetting experiments at the Solling, Germany. Water Air Soil Pollution 85:1251-1256

Evers FH, Moosmayer HU (1980) Relationships between site units, soil nutrient status and growth of spruce stands in a regional comparison. Forstw Cbl 99:137-140

Gross K, Pham-Nguyen T (1987) Einfluß von langfristigem konstanten Wassermangelstreß auf Netto-Photosynthese und das Wachstum junger Fichten (*Picea abies* [L.] Karst) und Douglasien (*Pseudotsuga menziesii* [Mirb.] Franco) im Freiland. Forstw Cbl 106:7-26

Gross K (1988) Nettophotosynthese, Produktion von Biomasse und Effektivität des Wasserverbrauchs junger Fichten und Douglasien unter dem Einfluß langfristig abgestufter Wasserversorgung im Freiland. Allg Forst- u Jagdztg159:230-239

Mehne Jakobs B (1995) The influence of magnesium deficiency on carbonhydrate concentration in Norway spruce (*Picea abies*) needles. Tree Physiology 15:577-584

Michael G, Tesche M, Feiler S, Ranft H, Bellmann C (1989) Physiologische Reaktionen der Fichte (*Picea abies*) auf komplexen SO_2- und Trockenstreß. Teil 2: Reaktionen trockenbelasteter Fichten auf nachfolgende SO_2-Einwirkung. Eur J For Path 19:293-304

Mohren GMJ, Ilvesniemi H, van Grinsven HJM (1993) Modelling effects of soil acidification on tree growth and nutrient status. Ecological Modelling 83:263-272Nilsson LO, Wiklund K (1992) Influence of nutrient and water stress on Norway spruce production in south Sweden - the role of air pollutants. Plant and Soil 147:215-265

Nilsson LO, Huttl RF (1997) Manipulation of conventional forest management practice to increase forest growth - results. For Ecol Manage 91:53-60

Nilsson LO, Wiklund K, Huttl RF (1993) Nutrient balance and P, K, Ca, Mg, S and B accumulation in a Norway spruce stand following ammonium sulphate application, fertilisation, irrigation, drought and N-free fertilisation. Plant and Soil 168:437-446

Rosengren-Brinck U, Nihlgård B (1995) Effects of nutritional status on drought resistance in Norway Spruce. Water Air Soil Pollution 85:1739-1744

Wiedemann E (1925) Zuwachsrückgang und Wuchsstockungen der Fichte in den mittleren und unteren Höhenlagen der sächsischen Staatsforsten. Tharandt

Wikstrom F, Ericsson T (1995) Allocation of mass in trees subject to nitrogen and magnesium limitation. Tree Physiology 15:339-344

2.3 Vitality and nutrient level of Norway spruce trees under changed environmental conditions

A. Dohrenbusch*, S. Jaehne*, H.W. Fritz**

* Institute of Silviculture, Göttingen University, Büsgenweg 1,
   D-37077 Göttingen, Germany, e-mail: adohren@gwdg.de
** Institute of Forest Botany, Büsgenweg 2,
   D-37077 Göttingen, Germany

Abstract

In the Solling mountains a 57-year-old Norway spruce forest stand has been manipulated by changing the quality (pre-industrial rain) and quantity (drought periods) over several years. The responses to artificially prepared, "pre-industrial" throughfall and to extended summer droughts with intensive rewetting are investigated concerning tree vitality and nutrient development In addition, biomass and element concentration of 100 trees have been investigated depending on needle age and position in the crown space.

The following results were found: The tree vitality in terms of needle losses and needle yellowing was improved during the observation period from 1992 to 1998. Compared to the control trees, only the trees of the drought experiment showed significantly worse vitality. Drought damages could be still observed three years after the drought experiment.

The concentrations of nitrogen, magnesium, phosphorus and potassium decrease with increasing needle age. The inverse trend could be proved for aluminium, calcium and manganese. The concentrations of the analysed elements decreased constantly from the upper to the lower parts of the crown. The variation of the element concentration inside the forest stand shows considerable differences between the elements: Nitrogen and phosphorus show a very low deviation, calcium and manganese have wide differences in concentration between the trees.

Key words: roof project, *Picea abies*, de-acidification, acid rain, clean rain, drought, crown vitality, needle loss, nutrient concentration, biomass

2.3.1 Material and Methods

From 1993 needle loss and yellowing was investigated for the trees of the roof and control areas. Also a visual assessment of vitality was carried out using a ranking scale of 1-10, in which completely healthy looking trees were ranked 1 and dead trees were ranked 10. This assessment of the crown was carried out from the crane by the same person every year. However, such an assessment is not scientific, especially comparisons between years have to be viewed with caution. More reliable is the comparison between the differently treated subplots monitored within one year. Between the years 1992-1996 in addition for almost all the trees photographs of the crown were taken from the same perspective, to optimize the annual damage assessment.

To determine the element concentrations in needles from 1992 onwards needle samples were taken from all trees of the roof and control sites. The sampling took place each year in September from the sun crown of the trees (6th whorl from the top). Only green branches were harvested, and only needles of the two most recent years were investigated. Needle samples from each year were dried to constant weight at 40 °C and milled. The needles were wet ashed under pressure (170 °C, 10 h) and elements determined by atomic absorption spectrophotometry (Perkin – Elmer 3030). Phosphate was determined colorimetrically. Nitrogen, carbon and sulphur were determined directly in the milled needles using a C/N analyser (Carlo-Erba).

In the pre-experimental phase of the project in 1989, twelve trees of the then unroofed areas were felled. On these trees in addition to differentiated biomass investigations the contents of elements was also determined (Dohrenbusch et al. 1993).

2.3.2 Results

2.3.2.1   Visual appearance of trees (vitality)

During the observation period the visual appearance (crown vitality) of the trees increased continually. This development was shown on all of the roofed areas (Fig. 2.3-1). Over the whole observation period the situation was different for the very healthy looking control trees outside the roofed areas. From 1993 to 1998 the appearance of these trees changed little, in every year they were assessed at an average value between 2 and 3.

The effect of experimental treatment was especially noticeable on the trees in the drought experiment. Their appearance was rated lowest at every assessment from 1993 onwards. This was due mainly to the high needle loss and the severe needle yellowing. The needle losses were from 1994 onwards significantly different compared to the control (Fig. 2.3-2).

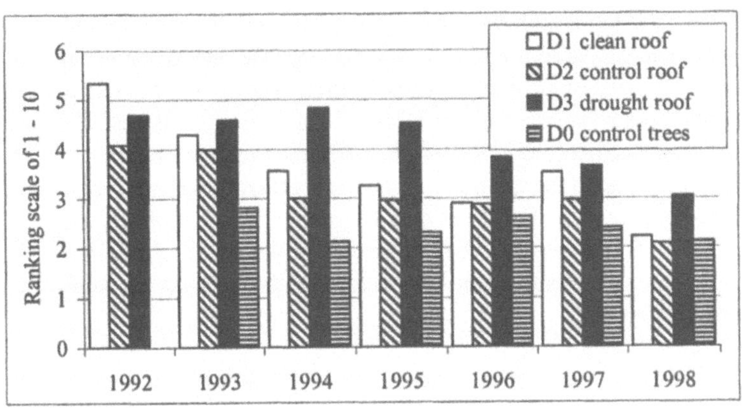

Fig. 2.3-1. Vitality (visual appearance) of the trees in the years 1992 to 1998.

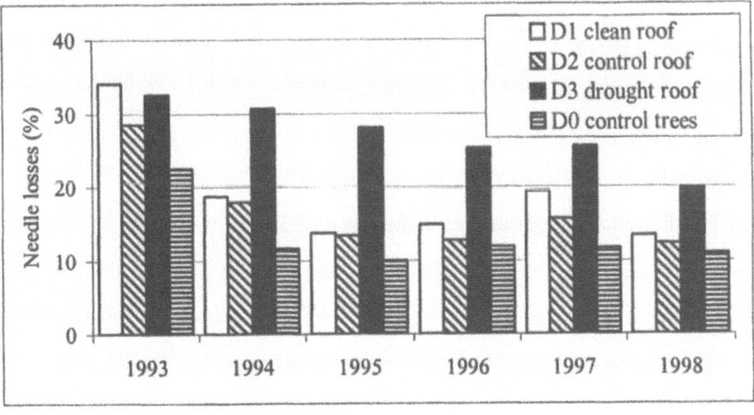

Fig. 2.3-2. Needle losses of the trees in the years 1993 to 1998.

As a result of the drought the spruce trees on the D3 site lost up to a third of their needles. In addition, in the years 1994-1995 the proportion of chlorotic (Fig. 2.3-3) needles was twice as high compared to the other areas. A regeneration of the trees was observed only two years after the end of the drought period. This was principally observed in the needle loss. In contrast, a complete regeneration of needle yellowing did not occur up to 1998. Here the region below the top portion of the crown (sub-top-dying) were particularly affected. In some trees in this crown area no new needles were formed three years after the end of the drought. It must be assumed that irreparable damage had occurred. The effect of the de-acidification experiment on the visual appearance of the tree is difficult to assess. The reason for this is that it was difficult to judge whether the improved appearance of the trees on the D1 site were due to the treatment or due to a general improvement. A summary of the visual appearance of all trees over the whole period of observation is shown in figure 2.3-4. This shows that the appearance of the tree on the D1 site increased by about 3 values. In contrast the appearance of the trees on the control site decreased by a value of 2.

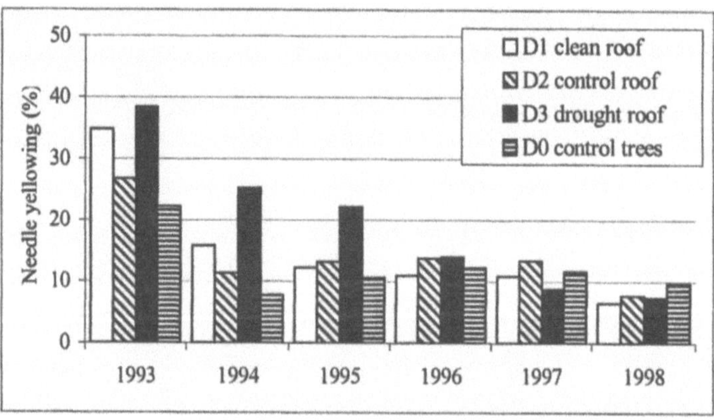

Fig. 2.3-3. Needle yellowing of the trees in the years 1993 to 1998.

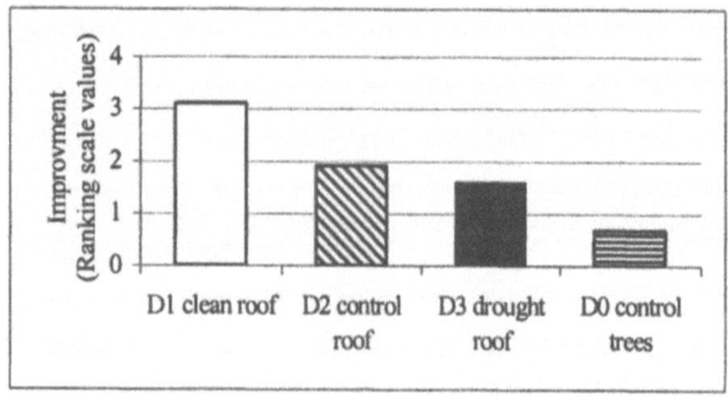

Fig. 2.3-4. Improvement of the vitality during the observation period (1993-1998).

The improvement of the trees on the de-acidification site was primarily due to a re-greening of the needles (Fig. 2.3-5). The mean value for chlorosis decreased over the observation period by ca. 28%. For the trees on the D2 site the decrease was only 17%. These values were statistically significantly different to the control treatment. In addition the average value for needle loss decreased by about 21% on the de-acidification site. On the other sites the corresponding values were 12 and 14%.

A conclusion was drawn that there is a high probability that the de-acidification treatment positively affected the optical appearance of the trees. This is supported particularly by the significant decrease in needle yellowing. In contrast, the drought experiment had negative effects on the number of needles. In some trees this high loss of needles was irreparable.

The differences between the treatments were much stronger if only the dominant and codominant trees of the different areas were considered. Thus environmental stresses, which are not lethal, will result in a strong shift in competition and subsequently in a change in the structure of the stand (more homogenous).

2.3.2.2   *Nutritional status of the trees*

Tables 2.3-1 to 2.3-3 show the average concentrations of the most important nutrients in the second year needles of all roofed trees (one-year-old needles were those needles that grew in the same year. Consequently needles of the previous year were labelled two year old, although they are only one and a half years old). Most of the mineral element concentrations showed no constant pattern of change over the five-year investigation period. Exceptions are the concentrations of nitrogen and potassium, which decreased on all sites from the beginning of the observations. By 1996 they had reached a level, which is considered to be deficient for spruce.

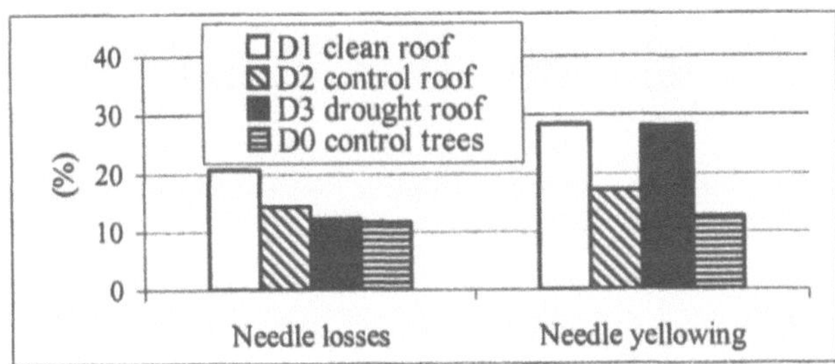

Fig. 2.3-5. Needle losses and needle yellowing during the observation period (1993-1998).

Table 2.3-1. Mineral element concentrations (mg g^{-1} dry mass, mean and standard deviation) in the two-year-old needles of the trees on the de-acidification plot.

Year	N	K	Mg	P	Ca	Mn	S
1992	15.7 ± 2.1	6.5 ± 1.3	0.4 ± 0.1	1.3 ± 0.2	2.6 ± 0.8	1.2 ± 0.4	$0.3 \pm 0,2$
1993	13.0 ± 1.1	4.6 ± 1.1	0.3 ± 0.1	1.0 ± 0.1	2.3 ± 0.7	1.2 ± 0.3	1.2 ± 0.2
1994	13.0 ± 1.1	4.8 ± 1.0	0.4 ± 0.2	0.9 ± 0.1	3.0 ± 1.2	1.5 ± 0.5	1.1 ± 0.2
1995	13.5 ± 1.2	4.4 ± 0.9	0.6 ± 0.2	1.0 ± 0.1	4.0 ± 1.8	1.8 ± 0.6	1.2 ± 0.2
1996	12.9 ± 1.1	3.8 ± 0.8	0.6 ± 0.2	1.1 ± 0.1	3.7 ± 1.3	1.7 ± 0.4	1.2 ± 0.2
1997	11.5 ± 1.0	4.6 ± 1.0	0.7 ± 0.2	0.9 ± 0.1	4.2 ± 1.0	1.7 ± 0.4	
1998	12.3 ± 1.3	4.7 ± 1.0	0.9 ± 0.2	1.1 ± 0.2	5.0 ± 1.1	2.3 ± 0.5	

Table 2.3-2. Mineral element concentrations (mg g^{-1} dry mass, mean and standard deviation) in the two-year-old needles of the trees on the control plot.

Year	N	K	Mg	P	Ca	Mn	S
1992	16.3 ± 2.3	6.1 ± 1.5	0.3 ± 0.1	1.2 ± 0.3	2.3 ± 1.1	1.0 ± 0.4	1.2 ± 0.2
1993	13.8 ± 1.8	4.7 ± 1.0	0.3 ± 0.1	1.0 ± 0.1	2.3 ± 1.0	1.1 ± 0.4	1.2 ± 0.2
1994	13.8 ± 1.8	5.1 ± 1.0	0.4 ± 0.1	1.0 ± 0.1	2.7 ± 1.0	1.3 ± 0.4	1.1 ± 0.1
1995	14.6 ± 1.4	5.2 ± 1.0	0.5 ± 0.2	1.0 ± 0.1	3.1 ± 1.2	1.4 ± 0.5	1.1 ± 0.2
1996	13.5 ± 1.4	4.3 ± 0.7	0.3 ± 0.1	1.0 ± 0.1	2.3 ± 0.9	1.0 ± 0.3	1.0 ± 0.1
1997	13.6 ± 1.9	6.0 ± 0.8	0.4 ± 0.1	1.0 ± 0.1	2.3 ± 1.0	1.3 ± 0.3	
1998	13.2 ± 1.4	4.9 ± 0.9	0.5 ± 0.2	1.1 ± 0.1	3.1 ± 1.1	1.4 ± 0.5	

Table 2.3-3. Mineral element concentrations (mg g^{-1} dry mass, mean and standard deviation) in the two-year-old needles of the trees on the drought plot.

Year	N	K	Mg	P	Ca	Mn	S
1992	14.8 ± 1.9	6.0 ± 1.0	0.3 ± 0.1	1.1 ± 0.1	1.7 ± 0.6	0.6 ± 0.2	1.3 ± 0.3
1993	13.0 ± 1.3	4.5 ± 0.8	0.3 ± 0.1	0.9 ± 0.1	1.8 ± 0.9	0.7 ± 0.2	1.2 ± 0.2
1994	13.0 ± 1.3	5.2 ± 0.8	0.3 ± 0.1	0.9 ± 0.1	2.0 ± 0.9	0.8 ± 0.3	1.1 ± 0.2
1995	14.2 ± 1.4	5.9 ± 1.1	0.4 ± 0.2	1.0 ± 0.1	2.5 ± 1.3	1.0 ± 0.4	1.2 ± 0.2
1996	14.3 ± 1.7	4.7 ± 0.9	0.5 ± 0.2	1.1 ± 0.2	3.5 ± 1.7	1.2 ± 0.5	1.2 ± 0.2
1997	13.5 ± 1.4	5.3 ± 0.9	0.5 ± 0.1	1.0 ± 0.1	3.5 ± 1.0	1.3 ± 0.3	
1998	13.5 ± 1.4	4.9 ± 0.8	0.6 ± 0.2	1.0 ± 0.1	4.0 ± 1.6	1.4 ± 0.5	

Obvious prolonged differences between the different treatments could not be seen. For the trees on the D3 site the levels of phosphorous and manganese were significantly reduced compared to the control treatment (D2) during the years with the strongest drought, 1993-1994. From 1995 in the needles of the trees on the de-acidification site a clear improvement was shown in the concentrations of calcium and magnesium. A possible explanation for this could be the artificial decrease in nitrogen input to the D1 site, resulting in a more balanced mineral nutrient content of the needles. This possibility was further investigated using N/Mg and N/Ca ratios. Both ratios can be used as indicators for balanced levels of nutrients in assimilating tissues. For trees on the roofed and unroofed control sites (D0 and D2) the same tendency can be seen. On the basis of samples from all control trees the ratios increased continually and reached values of 35-40 (N/Mg) and 5-6 (N/Ca) in 1995. This is due to not only the decreasing nitrogen content of the needles but also the increased Mg and Ca concentrations in the needles (Table 2.3-2). In the following years for both element ratios high values were found.

For the trees of the de-acidification treatment a completely different development was shown. Here the element ratios increased continuously to values of 18 (N/Mg) and 3 (N/Ca) by 1997. This trend was shown to be significantly different compared to the roofed control treatment from 1996 for the ratio of N/Mg and for 1995 for the N/Ca ratio. Consequently the trees on the de-acidification site had in relation to their nitrogen uptake a higher availability of other essential mineral nutrients. This advantage could be the reason for the increased growth of the dominating and co-dominating trees of the D1 site and their healthier appearance.

In the trees on the D3 site the highest values for the ratios of N/Mg and N/Ca were shown in the drought years. In 1995 the average Ca/N ratio value was significantly higher than that of the unroofed control, after which in the course of the regeneration the values decreased to less than those of both controlled treatments. This development suggests that drought stress negatively affects the element balance in the needles of the affected trees, although in the soil no acidification pushes could be induced.

The data presented as average values for the treatments mask that there are large differences between trees. For example for the two year old needles, for all trees the mean nitrogen concentration was 16.3 mg/g needle dry weight, the lowest single value was 11 mg g^{-1} and the highest 26 mg g^{-1}. The standard deviation of ±2.5 mg is by the above mentioned mean a variation coefficient of 15%. As a control this calculation was carried out again for the investigation of the following year 1993 (Table 2.3-4). Here the nitrogen concentration of the two year old needles had a variation coefficient of 12%, a somewhat smaller spread. In general these calculations showed that there are element specific differences in the value of the variation. Nitrogen and phosphorus with variation coefficients of under 20% are elements which are evenly distributed in the stand, or at least in a large part of the stand. Relatively homogenous values were also shown for sulphur, aluminium, potassium and iron whose variation coefficients were always between 20 and 30% and mostly between 20 and 25%. This is in contrast to the values found for magnesium which showed 30 to 35% deviations from the mean. Strong deviations were

                                          A. Dohrenbusch, S. Jaehne, H.W. Fritz

Table 2.3-4. Variation of the element concentrations in the needles of the spruce trees.

| Ele-ment | Needle age (Years) | Element concentration (mg g^{-1} dry matter) | | | Coeff. of variation (% Std.dev/mean) | | | |
| | | | | | Forest stand | | Single tree | |
		Mean and standard deviation	Min	Max	1992	1993	Strati-fied	All needle ages
N	1	16.3 ± 2.5	11	26	15	12	16	
	2	15.6 ± 2.1	11	22	14	12	8	17
P	1	1.39 ± 0.2	1.04	2.28	15	12	15	
	2	1.21 ± 0.21	0.81	2.15	17	13	17	38
S	1	1.18 ± 0.27	0.71	2.86	23	17		
	2	1.28 ± 0.26	0.86	2.55	20	22		
Al	1	0.12 ± 0.027	0.052	0.21	23	22	28	
	2	0.17 ± 0.039	0.077	0.27	23	24	33	63
K	1	6.7 ± 1.7	2.6	10.5	25	25	8	
	2	6.2 ± 1.4	2.7	10.5	21	22	18	30
Fe	1	0.083 ± 0.022	0.051	0.155	26	32	29	
	2	0.096 ± 0.020	0.059	0.164	21	22	14	53
Mg	1	0.46 ± 0.14	0.18	0.84	31	28	22	
	2	0.31 ± 0.11	0.16	0.61	35	27	34	39
Mn	1	0.97 ± 0.45	0.25	2.48	46	38	20	
	2	0.96 ± 0.41	0.27	1.99	43	42	21	35
Ca	1	2.12 ± 1.0	0.56	4.83	47	37	12	43

found for manganese, calcium, silicate and especially sodium. For these elements variation coefficients of over 40% were determined.

Figure 2.3-6 shows an example of the ranked distribution of phosphorus and calcium, two elements, which despite having similar mean concentration, have strongly different variation. A problem of the comparison of these elements is despite the use of a relative measure of dispersion such as the variation coefficient that the absolute contribution of the means is markedly different. For example, the mean nitrogen concentration of the two-year-old needles was 16.3 mg g^{-1}, more than 700 times higher than the mean sodium concentration of the same samples. It must be taken into consideration that at very low concentrations the accuracy of the measurements in the analysis used strongly affects the scattering of the data.

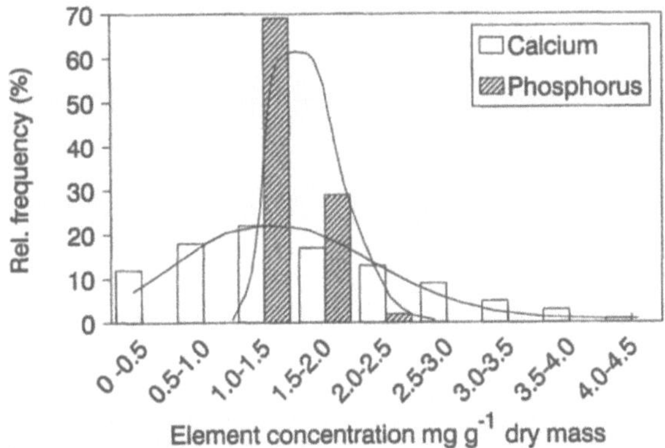

Fig. 2.3-6. Distribution of element concentrations of phosphorus and calcium of the total tree collective (N = 109 trees).

For the macro elements nitrogen, phosphorus, potassium and magnesium the variation between needles of the same age for all trees in a stand is not larger than the variation within a tree, based on samples of all needle ages and in all crown parts. Based on the data collected here only calcium showed a higher scattering at the stand level.

For the whole observation period it was not possible to show a clear association between the visual appearance of the trees and the element concentrations in the needles. This is as the heterogeneity of the element concentration within a tree is too high (Table 2.3-4, last two columns). Needle samples taken at a single site within the crown are not representative of a whole tree. Based on this, the reciprocal conclusion that the visually assessed vitality of the tree is associated with the element contents in the needles must be viewed with caution. This method for the identification and categorisation of deficiency symptoms as is commonly cited in the literature can only be used for heavily damaged trees. If only the absolute element concentrations are considered the correlation coefficients are seldom higher than 0.3**. Better results can be obtained if the ratios of nutrient elements are used. The closest correlation (r = 0.78**) could be shown for the 1995 values of the two-year-old needles between the N/Mg ratio and the percent chlorosis.

2.3.2.3   *Element storage*

The comparisons presented until now of the nutrient levels of the two-year-old needles from the sun crown were based entirely on element concentrations. For the estimation of the nutrient storage in the assimilating organs it was necessary to obtain data about the difference in element concentrations in the crown and, the dependency upon needle age. It was also necessary to obtain data about the compartmentation of the biomass i.e. the position and age. It could be shown that element concentrations varied considerably within single tree crowns. In addition cha-

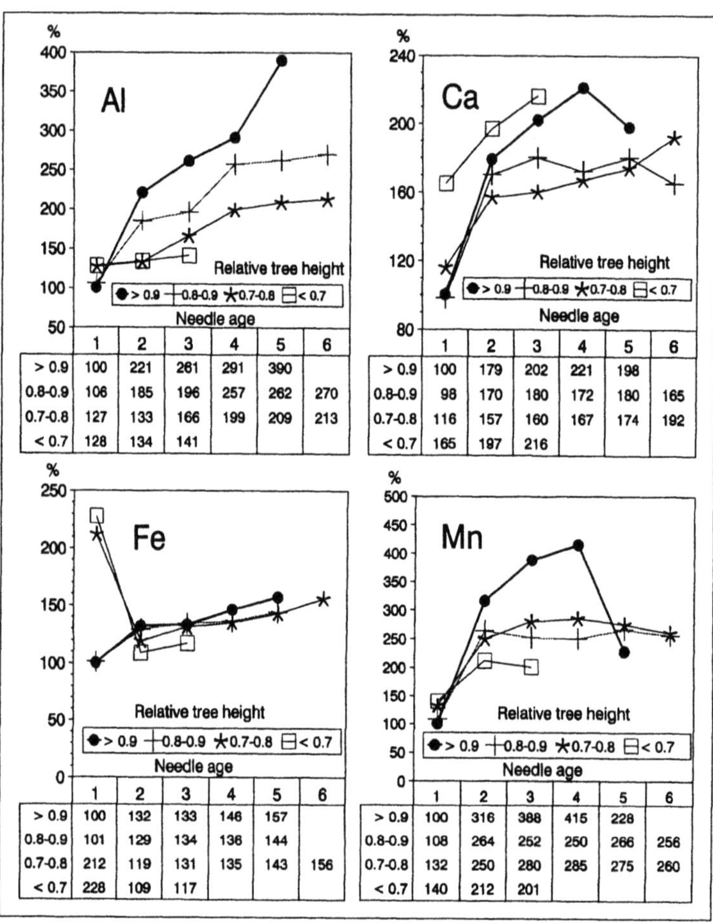

Fig. 2.3-7. Relative concentration of Al, Ca, Fe and Mn depending on the position in the crown (relative tree height) and needle age. The element concentration in young needles of the upper crown = 100%.

racteristic trends dependent upon needle age and relative tree height could be shown. This resulted in a division into two different basic types.

For nitrogen, phosphorus, magnesium and potassium the element concentration decreased as the needle age increased and the position on the tree decreased. Furthermore it could be shown that the differences between the light and shade crowns were greater the younger the needles were (Fig. 2.3-7). This phenomenon, which was only poorly shown for potassium, was accounted for by a strong age dependent decrease in the element concentration of the needles of the sun crown. Based on the highest value determined for the two-year-old needles in the sun crown the older needles in the lower crown had only half the maximum nitrogen value less than 40% of the phosphorus value and only 28% of the magnesium value. The greatest change in concentration due to age was shown for these elements between the one and two-year-old needles. Analyses of variants showed for nitrogen, phosphorus, and especially potassium significant** differences in the element concentrations of the one year old and all other needles. For magnesium two homologous groups could be identified which were significantly different from each other. One group composed of the first and second year old needles had high element concentrations, the other group was composed of the needle ages 3-6 with lower element concentrations.

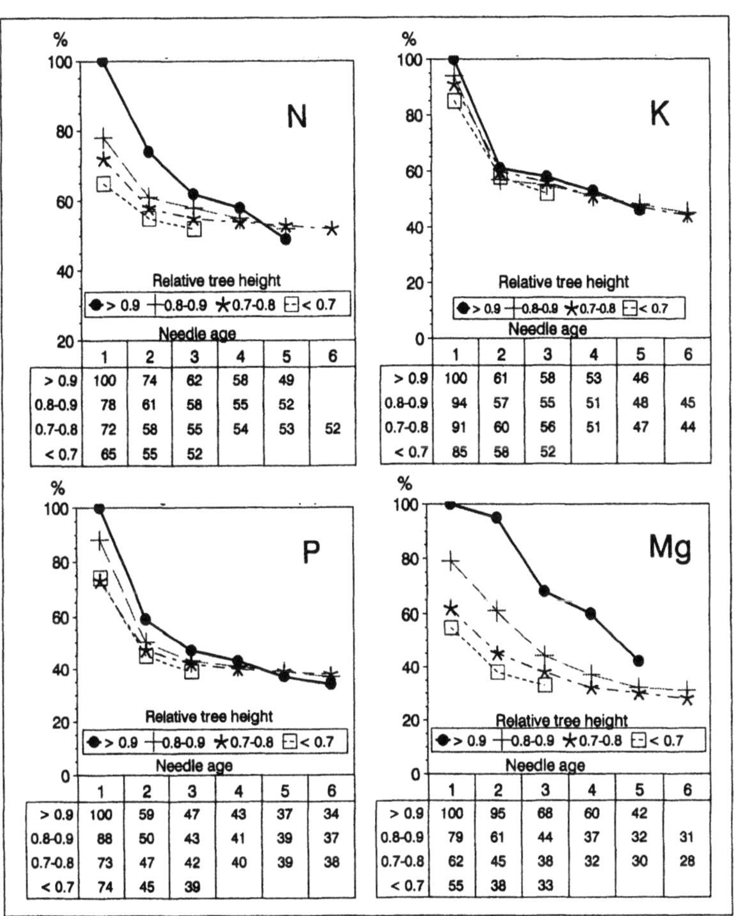

N

Needle age	1	2	3	4	5	6
> 0.9	100	74	62	58	49	
0.8-0.9	78	61	58	55	52	
0.7-0.8	72	58	55	54	53	52
< 0.7	65	55	52			

K

Needle age	1	2	3	4	5	6
> 0.9	100	61	58	53	46	
0.8-0.9	94	57	55	51	48	45
0.7-0.8	91	60	56	51	47	44
< 0.7	85	58	52			

P

Needle age	1	2	3	4	5	6
> 0.9	100	59	47	43	37	34
0.8-0.9	88	50	43	41	39	37
0.7-0.8	73	47	42	40	39	38
< 0.7	74	45	39			

Mg

Needle age	1	2	3	4	5	6
> 0.9	100	95	68	60	42	
0.8-0.9	79	61	44	37	32	31
0.7-0.8	62	45	38	32	30	28
< 0.7	55	38	33			

Fig. 2.3-8. Relative concentration of N, K, P and Mg depending on the position in the crown (relative tree height) and needle age. The element concentration in young needles of the upper crown = 100%.

In contrast to the element group described above a different pattern could be shown for calcium, manganese, aluminium and with restrictions iron (Fig 2.3-8). For these elements the concentration increased as needle age increased. For aluminium this could be shown over all needle ages. Additionally, the dependency on tree height as described above could also be shown. The age dependent increase in concentration was stronger the higher the position of the needles in the crown. Thus in the sun crown, five-year-old needles had a 4-fold higher aluminium concentration than one-year-old needles, whereas in the shade crown the difference was only 2-fold.

For manganese and calcium in the needles of the upper sun crown (relative tree height > 0.9) a trend to increased contents over the first 4 needles years was shown. For older needles a decrease in concentrations, which differed in degree was observed. In the middle and lower crown and obvious increase could only be seen in the one- and two-year-old needles. In needles years above this no or a slight decrease in concentration was shown as needle age increased.

To estimate the storage of elements in the assimilating parts on a tree or area basis, the concentration in the needles based on age and position must be combined with their respective dry weights. The required relationships, equations and distribution patterns are shown in Figures 2.3-9 to 2.3-11.

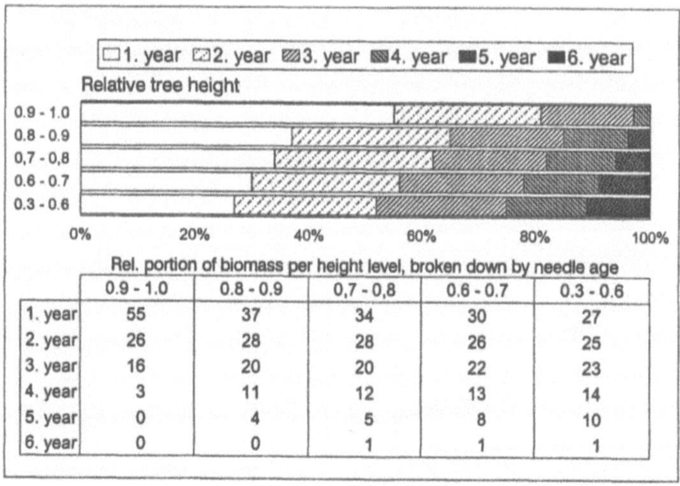

	0.9 - 1.0	0.8 - 0.9	0,7 - 0,8	0.6 - 0.7	0.3 - 0.6
1. year	55	37	34	30	27
2. year	26	28	28	26	25
3. year	16	20	20	22	23
4. year	3	11	12	13	14
5. year	0	4	5	8	10
6. year	0	0	1	1	1

Fig. 2.3-9. Distribution of needle mass, broken down by crown layer and needle age.

On the basis of this combined information it was possible to realistically calculate element storage. In addition, for each tree the yearly determined needle losses could be included in the model calculation of storages (Tab. 2.3-4). In 1994 the roof 1 area had a higher needle dry mass due to the higher stocking density. Roof 3 showed the lowest value after the intensive artificial drought period in summer 1994. The changes during the five year observation period from 1994 to 1998 showed marked differences. On the control area (roof 2) the needle dry mass increased by 12%, in comparison on the de-acidification sites (D1) a 14% increase was determined despite the higher initial value. Best improvement could be found at the drought site (D3) with an increase of 20% due to the recuperation after two years of drought.

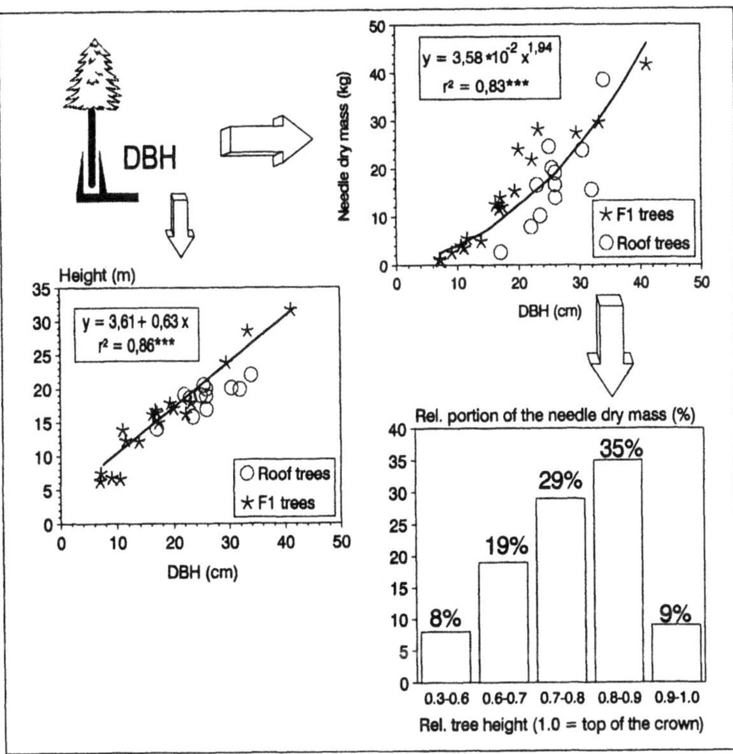

Fig. 2.3-10. Estimation of needle dry weight using the diameter at breast height (DBH) and distribution of biomass at the top layers of the crown.

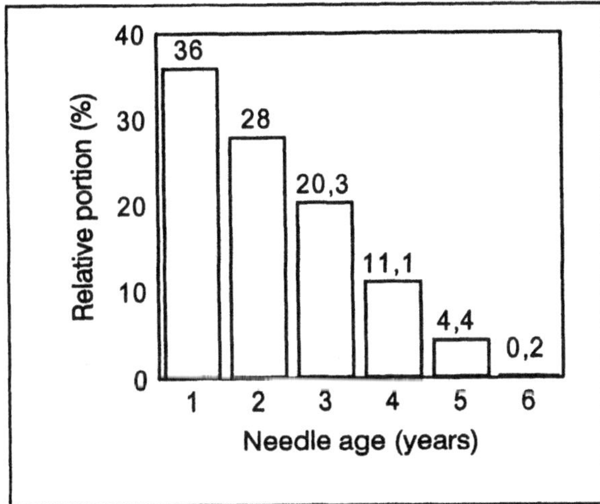

Fig. 2.3-11. Distribution of needle mass by age class of the average entire tree.

Tab. 2.3-5. Comparison of needle dry mass and element storage of the needle dry mass per hectare for the roof plots in 1994 and 1998 and change from 1994 to 1998 (%).

Roof	Year	Needle dry mass (Mass) and element storage of the needles (kg ha^{-1})									
		Mass	N	P	Mg	K	Ca	Mn	S	Al	Fe
1	1994	20520	166.1	10.7	4.9	61.9	95.1	57.8	22.5	4.1	2.3
	1998	23310	178.5	13.4	13.0	67.1	181.8	104.4	28.4	6.7	3.2
	%	114	107	125	265	108	191	181	126	163	139
2	1994	18000	154.4	9.8	4.0	55.8	75.7	46.0	19.2	4.0	1.8
	1998	20080	164.8	12.2	6.1	58.6	98.6	58.2	21.2	6.0	3.0
	%	112	107	114	153	105	130	127	110	150	167
3	1994	16023	132.7	8.3	3.1	53.0	50.3	26.1	17.4	4.3	1.7
	1998	19320	163.3	11.3	7.7	59.8	121.5	55.7	20.9	5.2	3.2
	%	120	123	136	248	113	242	213	120	121	188

If changes in single elements were observed it can be shown that there are some elements with strong changes at all experimental plots and others with moderate or no changes. Magnesium, calcium and manganese show a strong increase, especially on the drought plot D3 and the de-acidification site D1. The trees on the D3 site had an unusually strong increase in the calcium levels in the leaves and also a considerable increase in the storage of manganese. In general, the most obvious changes occured at this roof. The changes at the D1 roof as a result of cleaned precipitation were less obvious. The control roof D2 had the smallest changes.

2.3.3 Discussion

It was shown that both experimental variants, the de-acidification and drought experiment, have an effect on the vitality of the trees. The most obvious effect was manifest in the needle loss after several months of drought. Lenz and Schall (1991) confirm that drought stress is one of the main causes for unspecific needle loss. In the remaining needles of the trees of the drought roof significantly increased Manganese concentrations were determined. This result is also supported by investigations by Gärtner et al. (1990) showing a positive correlation between the Manganese concentrations and the needle losses in spruce trees of the same age in the Hessian uplands. Eichhorn (1987) considers the concentration of Manganese in the needles as an indicator for the efficiency of a root system. It appears that the increased Manganese content under drought conditions are related to a reduced fine root system resulting in a marked inhibition of the water and nutrient uptake.

Trees of the de-acidification experiment showed a considerable re-greening during the observation period. The crown area of these trees had showed yellowing symptoms at the beginning of the experiment. This regeneration is accompanied by a marked increase in the needle dry weight. It may be assumed that both incidents are due to the favourable nutrient condition of the spruce trees. This is also supported by the increased storage of Magnesium, Phosphor and Calcium found in the trees of the D1-site. The trees at this site are obviously able to compensate the reduced supply of Nitrogen and Potassium.

A large number of investigations have studied the relationship between yellowing and element contents of needles. Zöttl and Hüttl (1985) suggest that Potassium deficiency is one of the causes for forest decline in the uplands of the Alpes in south-west Germany. Zech and Popp (1983) related the decline of spruce and fir in north-east Bavaria to Manganese deficiency. Similar deficiencies in the nutrient supply according to Zöttl (1985) can be responsible for the needle yellowing in spruce trees in the higher uplands. Despite these findings, it appears to be difficult to relate visually identified deficiency symptoms directly to an insufficient supply of single elements. The de-acidification experiment at the D1-site has shown that the combination of the existing nutrient elements is of deciding importance for the vitality of trees. Deficiencies in single nutrients can only be compensated to a limited extent. The concept of the experiment, which was aimed at a change of the element inputs into the soil and a change in the assimilating organs at the same time, could be proved in the roof project. However, the expected changes for most elements did not occur until two or three years later. Hüttl and Liu (1986) determined a positive correlation between the element contents of one year old spruce needles and the soil only for Calcium and Phosphor. A possible explanation could be the immission loads or the leaching processes, respectively, which mainly result in losses of Potassium and Magnesium.

Independent of the experimental variants it could be shown that the distribution of nutrients in the needles shows marked element specific differences. The exceptionally high spatial variation, especially in Calcium, was also found by Knabe and Cousen (1988). In addition seasonal fluctuations in the nutrient condition of trees were determined (Rademacher 1994). Investigations of the spatial distribution within the crown showed characteristic trends in all cases. These are predominantly determined by the age of the needles and a vertical supply hierarchy. In all cases they resulted in distribution patterns, which can be explained by the element mobility in the leave tissue. The Calcium ions for example are not very mobile. They are fixed in the cell walls and can thus not be transported during the ageing process. As a result these elements are accumulated with increasing needle age (Knabe and Cousen 1988). By contrast, Manganese and Potassium are hardly inhibited in their mobility and can be transported more easily from ageing needles. This theory explains the decrease of Magnesium and Potassium with increasing needle age. The same applies for Nitrogen and Phosphor concentrations. Comparable results were obtained by Fritz (1990), Leibundgut (1985) and Reemtsma (1986). It must be considered that the element concentrations are usually based on the dry mass of the needles. As this increases with increasing needle age (Schulze et al. 1977) changes in the element concentrations (independent of the absolute amount of elements) can be reinforced or compensated, respectively.

Acknowledgements

The study received financial support from the German Federal Ministry of Research and Technology (BMBF, 325-7291 OEF 2019) and from the State of Lower Saxony. Furthermore, we want to give our special thanks to the technicians G. Brauer, A. Nannen, K.-H. Obal, D. Pryor, U. Schmidt, C. Suner and H. Zimmer for their collaboration and O. Godbold for improving the English.

References

Eichhorn J (1987) Vergleichende Untersuchungen von Feinwurzelsystemen bei unterschiedlich geschädigten Altfichten (*Picea abies* Karst.). Forsch Ber Hess Forstl Versuchsanst Hann Münden 3

Fritz HW (1990) Methodische Untersuchungen zur Bestimmung von Biomasse und assimilierender Oberfläche in einem 53jährigen Fichtenbestand im Solling. Diplomarbeit Fak f Forstwissenschaften u Waldökologie Univ Göttingen

Gärtner EJ, Urfer W, Eichhorn J, Grabowski H, Huss H (1990) Mangan - ein Bioindikator für den derzeitigen Schadzustand mittelalter Fichten in Hessen. Forstarchiv 61:229-233

Hüttl RF, Liu J (1986) Beziehungen zwischen Elementgehalten in Nadeln und chemischen Bodenparametern von geschädigten Fichtenökosystemen. Kolloquium d Oberrheinschen Universität Umweltforschung in der Region 27.-28.6.86 Straßburg.

Knabe W, Cousen G (1988) Regionale Verteilung einiger Nähr- und Schadstoffgehalte in Fichtennadeln. Schr d Bundesministers f Ernährung, Landwirtschaft u Forsten, Reihe A, H. 360, Landwirtschaftsverlag., Münster-Hiltrup

Leibundgut H (1985) Über den Mineralstoffgehalt frischer Fichten- und Tannennadeln. Schweiz Z Forstw 136:49-53

Lenz R, Schall P (1991) Risikoarten zu Schlüsselprozessen der neuartigen Waldschäden. Allg Forstz 15:756-761

Rademacher P (1994) Qualitative und quantitative Erfassung der Stoffflüsse und der Elementvorräte in Waldbäumen und deren Umwelt. Ber Forschungszentrum Waldökosysteme, Göttingen, Reihe B, Bd 37

Reemtsma JB (1986) Der Magnesium-Gehalt von Nadeln niedersächsischer Fichtenbestände und seine Beurteilung. Allg Forst- u Jagdztg 157:196-200

Schulze ED, Fuchs MI, Fuchs M (1977) Spacial distribution of photosynthetic capacity an performance in mountain spruce forest of northern Germany, I. Biomass distribution and daily CO_2 uptake in different crown layers. Oecologia 29:43-61

Zech W, Popp E (1983) Magnesiummangel, einer der Gründe für das Fichten- und Tannensterben in NO-Bayern. Forstwiss Cbl 102:50-55

Zöttl HW, Hüttl R (1985) Schadsymptome und Ernährungszustand von Fichtenbeständen im südwestdeutschen Alpenvorland. Allg Forstz 40:197-199

Zöttl HW (1985) Waldschäden und Nährelementversorgung. Düsseldorfer Geobot Kolloquium März 1985, Düsseldorf:31-41

3 Effects of management practices on ecosystem processes in European beech forests

3.1 Effects of group selection and liming on nutrient cycling in an European beech forest on acidic soil

N. Bartsch*, J. Bauhus, T. Vor***

* Institute of Silviculture, Göttingen University, Büsgenweg 1,
D-37077 Göttingen, Germany, e-mail: n.bartsch@forst.uni-goettingen.de,
tvor@gwdg.de
** School of Resources, Environment & Society, The Australian National
University, Canberra, ACT 0200, Australia,
e-mail: juergen.bauhus@anu.edu.au

Abstract

Nutrient cycling in forest gaps has received little attention until now, although gap regeneration is important for natural dynamics and forest management practices in temperate forests. Gaps of 30 m diameter were cut in a mature European beech forest (*Fagus sylvatica* L.) in the Solling Hills in Northwest Germany. The forest has been destablished through prolonged atmospheric pollutants. Deposition, plant uptake, nitrogen mineralization, seepage water chemistry and element losses were compared between limed and untreated gaps and the mature beech stand. The most obvious changes in the element cycle in gaps occurred in the output fluxes of nitrate, cations, heavy metals and laughing gas. The seepage water concentrations of nitrate and cadmium in the unlimed gaps exceeded drinking water thresholds. The major difference in seepage water concentrations of nutrients between gaps was caused by the development of the herbaceous vegetation in the limed gaps. Gap creation did not increase N mineralization as repeatedly reported for large-scale clear cuts.

The results stress the importance of effective and early coupling of decomposition and nutrient uptake by new vegetation after tree removal. Where soil acidity limits rapid revegetation, liming promote the establishment of herbaceous vegetation and thus the nutrient retention through plant biomass production.

Key words: *Fagus sylvatica*, deposition, element fluxes, gap dynamic, ground vegetation, input-output budgets, litter, nitrogen mineralization, soil fauna, soil solution, tree regeneration

3.1.1 Introduction

Gap disturbance

Disturbances have major implications for community structure and dynamics, as well as for ecosystem properties (Attiwill 1994). One of the most fundamental parameters of any disturbance is its extent. The extent of the physical disruption not only dictates the amount of physical space available for colonisation, but also greatly influences the types and amounts of resources. Light and soil moisture profiles, soil nutrient content, and factors that modify the utilisation of these resources, such as air and soil temperature, are dependent on the extent of damage to the vegetation or removal. Larger disturbances often create environments, where extreme changes occur both at the level of resources and in factors such as light, temperature and wind speed, which regulate resource utilisation (McConnaughay and Bazzaz 1987). In temperate and tropical forests, small-scale disturbances ('minor disturbances or perturbations' sensu Oliver 1981) at the scale of individual trees are of major importance to natural succession, while large-scale disturbances at a landscape level are rare (Runkle 1985, Platt and Strong 1989, Denslow and Spies 1990, Röhrig 1991, Attiwill 1994).

A gap is a product of a small-scale disturbance. The term comes from the description of the forest cycle by Watt (1947), in which he defines the gap phase of the cycle as that "to which regeneration is confined because it is excluded from other phases" (Runkle 1992). In the following, gap disturbance is viewed as a change in stand structure caused by the death of a large branch, a single tree, or a small group of trees. The largest gap is created by the death of 10 canopy trees or has a ratio of canopy height to gap diameter equal to 1.0 (Runkle 1992). Most definitions of gaps restrict them to the areas directly under the canopy opening (canopy gap after Runkle 1982). The openings in the forest canopy may range from < 25 m^2 to about 0.1 ha (Runkle 1982). The variability in the size of canopy gaps observed in forest communities is the result of two processes: (1) their creation after tree death, and (2) their disappearance as a consequence of the growth of suppressed saplings and newly established tree regeneration, and the expansion of crowns of surrounding trees (Holeksa 1993).

Tree regeneration caused by small-scale disturbance is common in forest ecosystems composed of long-living shade tree species. In old-growth hardwood forests in Eastern North America, the gap formation rate (fraction of ground area converted to new gaps per year or fraction of canopy trees or canopy area dying per year, Runkle 1989) differs between $< 0.4\%$ (Barden 1981) and 2% (Runkle 1982, 1985). The gap size ranges typically from 50 to 200 m^2 for single tree falls, to up to 300-500 m^2 for multiple tree falls (Canham et al. 1990, Attiwill 1994). For deciduous forests of the temperate zones of Japan (Yamamoto 1989, 1996) and for *Nothofagus*-forests in New Zealand (Runkle et al. 1995) and Argentina (Rebertus and Veblen 1993) similar gap dynamics have been described.

Gaps in European beech forest ecosystems

Without anthropogenic influence European beech (*Fagus sylvatica* L.) forest communities would be dominating even the high uplands in Central Europe, where small-scale disturbances play an important role in forest dynamics (Bengtsson et al. 2000). Information about gap dynamics and gap phase regeneration have been gained from native beech forests of the Carpathian mountains (Slovakia, Poland) and the Balkan (Slovenia, Albania). Investigations by Korpel (1995) have shown that the regeneration in native beech forests in eastern Slovakia do not exceed 2000 m^2, in most cases they are much smaller. The small gaps are usually not increased by sun scald or windthrow. In a Carpathian beech forest in Poland, Holeska (1993) counted 83 gaps ranging from 20 to 2330 m^2 in an investigated area of 30 ha. Most gaps were between 50 and 100 m across with an average area of 179 m^2. Diaci and Boncina (1997) referring to two Slovenian montane beech/fir primeval forests gave similar accounts. Under favourable site conditions gaps of less than 500 m^2 were mostly found, the largest gap measured 650 m^2. The gaps account for 1.2% of the primeval forest area. Under unfavourable site conditions gaps were larger (up to 2000 m^2) and the proportion of gaps was higher (2.2%). The gap area in Albanian native beech forests lies between 3.3% and 6.6% and average gap size between 60 m^2 and 74 m^2 (Tabaku and Meyer 1999). Further information about gap sizes are available from investigations that have been carried out for at least several decades in unmanaged forest nature reserves. For a beech stand in Northeast Germany that has been left unmanaged since 1850 and is in the disintegration phase, gaps of up to 400 m^2 were determined (Knapp and Jeschke 1991).

The results show that the regeneration in unmanaged beech forests mainly occurs in small gaps, and that a successional growth of woody vegetation and pioneer trees is less probable. A change of tree species to pioneer trees and an increase in species diversity as Remmert (1991) has claimed to occur normally in beech forests in Central Europe, can not be supported by the investigations carried out in unmanaged forests (Schmidt 1998). Moreover, it seems that regeneration after the collapse of the old beech trees is taken over by the shade tolerant species, which were already in the regeneration phase, without a separate forest community being established. Depending on the site conditions this regeneration type is often dominated by beech (Watt 1947, Röhrig and Bartsch 1992, Korpel 1995, Schmidt 1998). Broad-leaved tree species (*Acer, Fraxinus, Prunus, Ulmus, Tilia*) only participate in this process on nutrient rich, fresh soils. On montane sites spruce (*Picea abies* L.) and fir (*Abies alba* L.) also participate. This indicates that the the tree species composition in unmanaged European forests has a high degree of stability.

Regeneration of European beech

In managed forests, it is a priority to avoid substantial disturbance in the stands using appropriate silvicultural methods. In contrast to natural forests, the dominating factor for renewal of managed forests is replacement, not natural ageing and subsequent death. On the contrary, the aim is to harvest and renew the trees at their highest value, before symptoms of age begin to appear.

For no other tree species in Central Europe is natural regeneration so important in practical silviculture as for beech. Since it colonized central Europe beech has dominated most forest communities completely, in many others it plays an important role as a mixed species. This is due to its high shade tolerance in regeneration and young growth phases, its long sustained growth and its ability to spread its crown up to a high age. These properties put beech in an advantage over less shade tolerant tree species at many sites. The main silvicultural method for regeneration of beech is shelterwood cutting. Using this method the whole stand is simultaneously regenerated. In doing so the crown cover is evenly more and more reduced until it is completely removed. This method favours the formation of structurally poor stands.

In face of unprecedented large-scale disturbances in Central Europe over the past 20 years caused by windthrow, fire and a new type of forest damage, a fundamental change in the assessment and silvicultural methods used in forests has been initiated. This has resulted in forest management concepts, which aim at a long term restructuring of forests by adjusting artificial species composition and structure of forests towards that of natural forest communities. The basic measures implemented in natural forest management include the preference for natural regeneration, an increase of deciduous and mixed forest, increase in structural diversity and target diameter felling as single stems or groups (Otto 1995).

Constant tree removal by thinning and the maintenance of a continuous cover only promotes shade tolerant plants. Regeneration in a local area or over a long time period and in scattered, ever-increasing gaps (group selection) is overall a more efficient concept to promote biodiversity (Scherzinger 1996, Schütz 1998). The size of the gaps plays a very important role in this method. In temperate hardwood forests of North America, high species diversity can only be established with interspersed multiple-tree gaps of at least 500 m^2 (Lorimer 1989). Hibbs (1982) even suggests gap sizes of more than a diameter of 50 m. In Central Europe gap sizes of 500 m^2 are considered to be needed if light favouring tree species (e.g. *Quercus*) are to be introduced in stands with dominating shade tolerant tree species (Lüpke 1998).

Objectives of the case study

Gap research has mainly been descriptive and has focused on phenological processes. Many studies in forest stands of different regions have focused on tree regeneration and the development of ground vegetation in gaps (e.g. Mosandl 1984, Lüpke 1987, Bazzaz and Wayne 1994, Kuuluvainen 1994, Yamamoto 1996, Ban et al. 1998) and the effect of changed microclimatic conditions, such as radiation, temperature and moisture (Luft 1973, Minckler et al. 1973, Mosandl 1984, Bazzaz and Wayne 1994). Simulation of regeneration, growth, and mortality in gaps has been the basis of many forest succession models (Botkin 1993, Shugart and Smith 1996, Waring and Running 1998). Few studies have investigated the effects of gap size canopy removal on soil processes (Mladenoff 1987, Parsons et al. 1994). The changes in nutrient cycling, which occur after canopy removal, reorganisation and the subsequent aggradation phase in gaps, have not been sufficiently studied. These changes will primarily result from the removal of the canopy, the interrupted root water and nutrient uptake and the input of organic matter from the biomass of dead trees in natural gaps. Canopy removal creates distinct microclimatic gradients from the surrounding stand into the gap (Collins and Pickett 1988, Geiger et al. 1995), which influence water balance and nutrient cycling.

As part of the International Biological Program (IBP), ecosystem research in a mature beech forest ecosystem in the Solling, a mountain area in Northwest Germany, has been continuous since 1966 and thus represents one of the longest continuous measurements of element cycling world-wide (Ellenberg et al 1986, Matzner 1989). The measurements revealed high rates of deposition of various air pollutants, especially sulfuric acid, and detrimental effects of acid deposition on soils and forest ecosystems (Ulrich et al. 1979, Ulrich 1994).

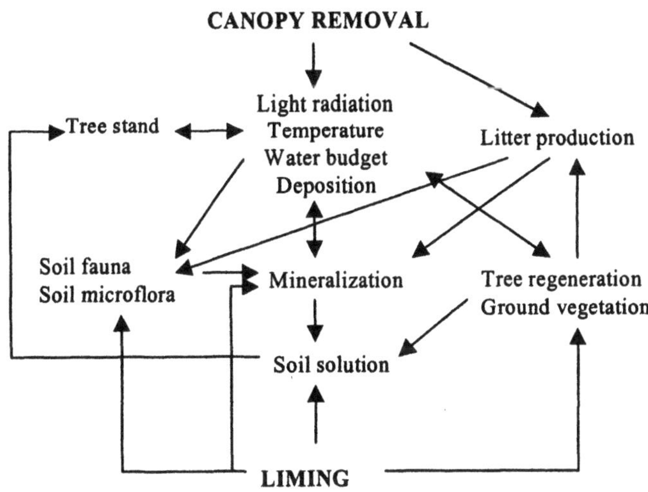

Fig. 3.1-1. Potential effects of canopy removal and liming on selected ecosystem processes and compartments.

To investigate the effects of small scale disturbances and stand regeneration on decomposition processes and water and element fluxes, gaps were cut in an area of the Solling beech stand in 1989. In the past it was shown that natural beech regeneration on acidic sites like the Solling area was more successful when the soil was limed (Gehrmann 1984). Therefore this study was also conducted in limed gaps.

The purpose of the case study was to describe water and element fluxes in changing ecosystem compartments as well as the relevant ecological factors during succession in beech forest gaps (Fig. 3.1-1). The results will facilitate recommendations for management practices in accordance with ecosystem processes.

3.1.2 Study area

The study area is located on the plateau of the Solling Hills in Northwest Germany (Lower Saxony) at about 500 m elevation. The Triassic red sandstone of the Solling plateau is overlain with a 40 to 80 cm thick layer of loess. The soil type is a silty loam (20% sand, 60% silt, and 20% clay). The mineral soil is heavily acidified and within the aluminum buffer range (Table 3.1-1). This intensively researched IBP-site is described in more detail by Ellenberg et al. (1986) and Matzner (1989).

Table 3.1-1. Site and stand characteristics

Latitude	51°46′ N
Longitude	09°34′ E
Inclination	3° (SE)
Elevation above sea level (m)	495
Long-term average precipitation (mm per year)	1032
Annual mean air temperature (°C)	6.4
Soil parent material	weathered Triassic red sandstone covered by loess sediments
Humus form	typical mor
Soil type (FAO)	dystric Cambisol
Soil pH(CaCl$_2$)	2.9-3.6 (0-10 cm)
Effective cation exchange capacity (μmol$_c$ g^{-1})	138 (0-5 cm)-44 (30-40 cm)
Base saturation of the CEC[1] (%)	< 5%
Al of the CEC[1] (%)	56% (0-5 cm)-95%(30-40 cm)
Tree species	*Fagus sylvatica* L.
Tree age (in 1989) (years)	145
Stand density (trees ha^{-1})	200
Basal area (m^2)	32.5
Timber volume (m^3)	488

[1]CEC, effective cation exchange capacity

The vegetation type of the climax forest is *Luzulo-Fagetum typicum*. Typically this is strongly dominated by European beech (*Fagus sylvatica* L.). Beech forms a dense crown cover, which prevents the development of a shrub layer and restricts regeneration to gaps. The studied beech stand developed from natural regeneration (land-use history see Ellenberg et al. 1986). At the beginning of the experiment in 1989 the stand was about 145 years old (yield surveys see Table 3.1-1). The ground vegetation layer in the mature beech stand is extremely sparse (about 3% cover) and is dominated by *Luzula luzuloides, Oxalis acetosella, Carex pilulifera* and the moss species *Polytrichum formosum* and *Dicranella heteromalla*.

3.1.3 Experimental design and methods

Four gaps of 30 m diameter were cut and the total research area of 8 hectare was fenced in October 1989. The diameter of 30 m was chosen in order to obtain a ratio of the gap size to the stand height of 1. With other ratios the microclimatic conditions approach the field conditions increasing the danger of radiation frost (Geiger et al. 1995). Smaller gap sizes would not allow the differentiation of the microclimatic conditions, which is necessary for the causal analysis of decomposition processes and plant growth. In the studied beech stand, a gap with a diameter of about 30 m has already been created in 1972 by windthrow (Stickan 1993). Gaps of about 700 m^2 are also found in the stands of eastern European native beech forests, however, smaller gaps occur more frequently.

Stems and branches were removed from the gaps. Two gaps, including a 20 m wide belt in the surrounding stand, were limed with 3 t ha^{-1} of fine dolomite (10.6 kmol$_c$ kg^{-1} $CaCO_3$, 11 kmol$_c$ kg^{-1} $MgCO_3$) to improve growth conditions for beech seedlings. Lime was slightly mixed with the organic horizons by superficial scarification. The remaining two gaps were untreated (Bauhus and Bartsch 1995).

Precipitation, litter fall and air and soil temperature were monitored along north-south and east-west transects which crossed the gaps and extended into the adjacent mature stand (Fig. 3.1-2). Deposition was determined using bulk precipitation gauges on a tower above the forest (open field precipitation), under the canopy and in the gap and stem flow collectors in the stand (Ellenberg et al 1986, Bauhus 1994). For a large-scale characterisation of the radiation conditions in the gaps using methods developed by Wagner (1994, 1998), hemispheric 'fish eye' photographs were taken for the whole area. Radiation conditions above the ground vegetation within the gaps and in the surrounding stand (DIFFSF = diffuse site factor according to the procedures of Anderson 1964) were expressed as a percentage of the conditions occurring above the crowns of the mature stand (relative light intensity).

Soil solutions were collected with ceramic suction lysimeter cups (KPM, Berlin Type 80) in three 10 m x 10 m-plots (centre, northern edge, southern edge = stand, Fig. 3.1-2) of one unlimed gap and in the centre of one limed gap. Lysimeters were installed at 0, 5, 10, 15 and 80 cm depth using three lysimeters at each depth per plot. The chemical composition of the soil solution at 80 cm depth was considered to be the output concentration of seepage water as it was below the rooting zone of the beech trees. Soil solution and precipitation samples were collected fortnightly and combined proportionally to provide monthly samples. The soil water potential was measured by tensiometry at 15 minute intervals using an automated field system (Schmidt 1991). A water balance model was employed to calculate monthly seepage water fluxes (Schmidt et al. 1995). The analytical methods used are described by König and Fortmann (1996). From March 1991 to May 1992 and from May 1996 to May 1997, C and N microbial biomass and nitrogen mineralization were monitored simultaneously at the centre of a limed and an unlimed gap, and along two transects in the stand (Bauhus 1996, Vor 1999). Soil N_2O emissions were measured using an automated chamber method (Loftfield et al. 1992, Brumme 1995).

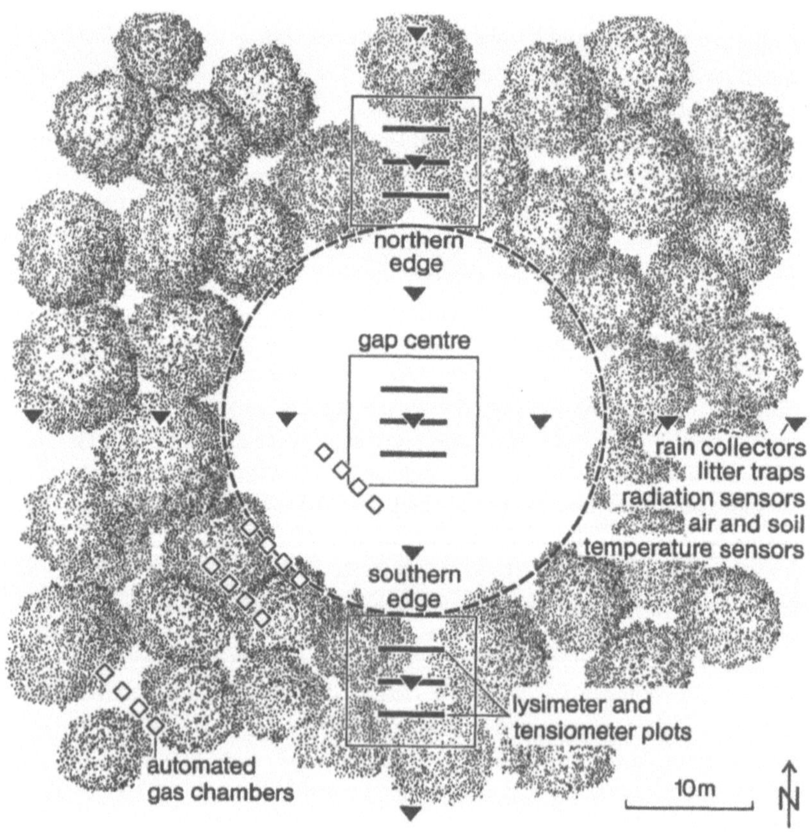

Fig. 3.1-2. Instrumentation and distribution of plots in a gap of the mature beech stand.

In March 1992 and May 1996 litterbags (10 cm x 10 cm, mesh width 1.8 mm) with roots and leaves of beech and *Epilobium angustifolium* were laid out for the assessment of the decomposer activity in the upper soil. For the determination of weight loss, the litterbags were removed from the soil after 4, 8, 12 and 16 months (Bauhus, 1994; Vor 1999).

To estimate the growth of ground vegetation and tree regeneration, in each 10 m x 10 m-plot (Fig. 3.1-2) 32 sample plots of 1 m x 1 m were established. The biomass of ground vegetation and tree regeneration was determined in August 1993 and 1996 at the point of maximum seasonal biomass accumulation. Biomass was harvested by clipping all above ground vegetation from 10 to 20 randomly selected 0.25 m²-plots in the gap centre and at the northern and southern edge of the gaps. Below-ground biomass was sampled by soil coring (8 cm diameter, 20 cm depth). Two cores were collected from each plot in which aboveground biomass had been harvested. Soil was separated from the roots by the floatation method, and the roots were collected on a sieve (<1 mm mesh) (Bauhus and Bartsch 1996).

3.1.4 Microclima and soil water

The effects of clear cuts on the stand climate have been described in detail for several central European forest ecosystems; for group selection felling in fir-beech stands (Luft 1973), for gaps and shelterwood cutting in beech stands (Lüpke 1982), for gaps in mountainous mixed forests (Mosandl 1984, Röhrig and Bartsch 1992, Geiger et al. 1995). Therefore climatic factors which are unimportant for decomposition processes and the growth of vegetation in the gaps are only briefly mentioned here.

Based on hemispheric photographs the relative light intensity (DIFFSF) at 1.3 m height above the ground vegetation was determined (Fig. 3.1-3, data from 1996). The radiation conditions near the gap centres reached about 50% of the radiation above the crowns of the stand. Such radiation conditions were generally judged to be favourable for the development of beech regeneration (e.g. Lüpke 1982, Mosandl 1984). Only a small proportion of the radiation measured above the ground vegetation was available for the beech regeneration growing below the ground vegetation in the limed gaps (see subsequent chapter). At the centre of the limed gaps (95% percent cover of ground vegetation) radiation was measured directly under the herbaceous ground vegetation (quantum sensor LI-190SA LICOR Instruments). In 1991, on average only 3% (0.5-9.0%) of the above canopy radiation was measured at ground level of the limed gaps. These are low light levels for the developing beech regeneration.

The daily fluctuation in temperature in the upper soil layers (monitored at 2, 5, 8 and 15 cm soil depth) was higher in the gaps than in the stand. However, only at the northern edge the daily mean- and annual mean temperatures were slightly increased compared to the stand. The daily maxima were also highest below the northern edge due to a higher amount of direct sunlight. In summer, the temperatures in the stand at 2 cm soil depth were more than 15 °C lower than at the same soil depth at the northern edge of the gap (Fig. 3.1-4). In the gap centre the temperatures at 2 cm soil depth on some occasions was about 5 °C higher than in the stand. The daily minima of all sites differed only slightly. Relative temperature differences between sites were similar but less pronounced at 15 cm soil depth. The maximum temperature difference between 2 and 15 cm soil depth were up to 12 °C in summer and up to –2 °C in winter.

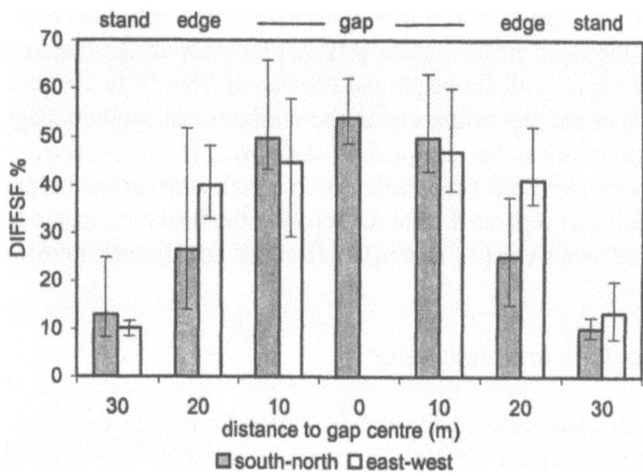

Fig. 3.1-3. Distribution of relative light intensity (DIFFSF) within gaps and in the surrounding mature beech stand on a south-north and a west-east transect (mean and min/max of DIFFSF).

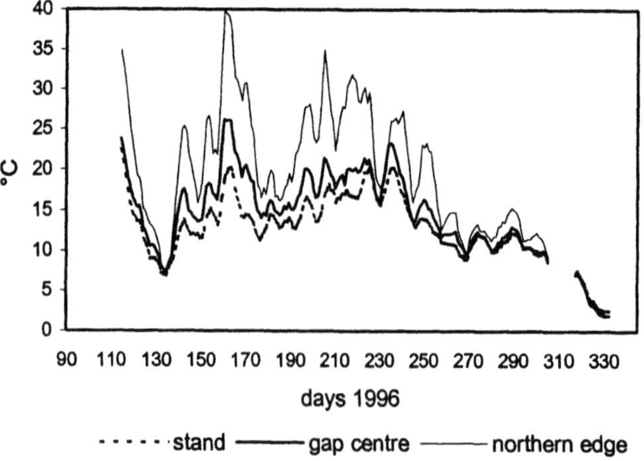

Fig. 3.1-4. Maximum daily soil temperature (°C) at 2 cm soil depth in the stand, the gap centre and the northern edge of the surrounding stand (moving average of five subsequent days).

Soil water tension changed dramatically after removal of the trees (Fig. 3.1-5). Whereas the soil in the stand became dry in late summer, soil water tension in gaps remained low throughout the year. The herbaceous vegetation in limed gaps had only a small influence on the soil moisture in the first five years after gap formation. Even at 5 cm depth the soil water potential did not exceed 150 hPa. After the fifth experimental year the soil water potential of the limed gap with lush ground vegetation has been about 100 hPa above the values of the unlimed gap.

3.1.5 Nutrient storage in ground vegetation and beech regeneration

In 1989, the gaps were created in parts of the stand which were predominantly free of vegetation (percentage vegetation cover less than 5%). A significant vegetation establishment in non-limed gaps did not occur until the 4th vegetation period after formation of the gaps (Fig. 3.1-6). This process was initially attributable to beech regeneration, and since the 5th vegetation period also to herbaceous species.

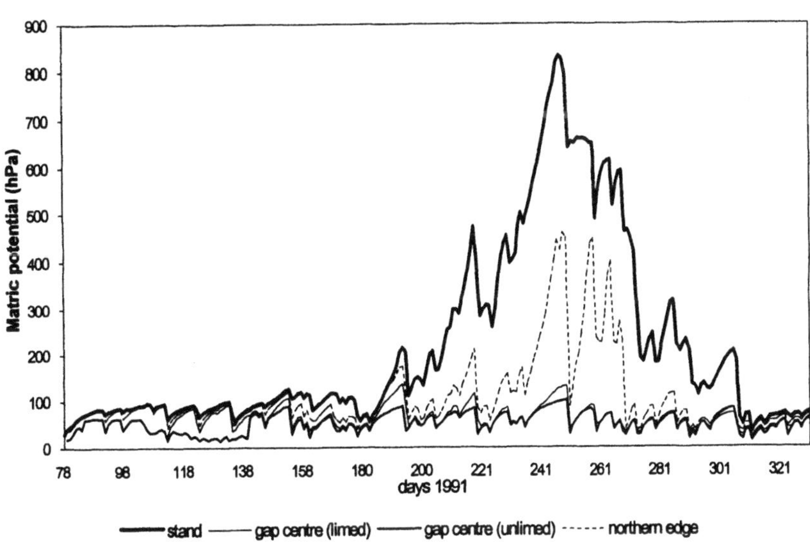

Fig. 3.1-5. Soil water potential (hPa) at 15 cm depth during 1991 in the stand, at the edge of the surrounding stand, in the centre of the unlimed gap and in the centre of the limed gap.

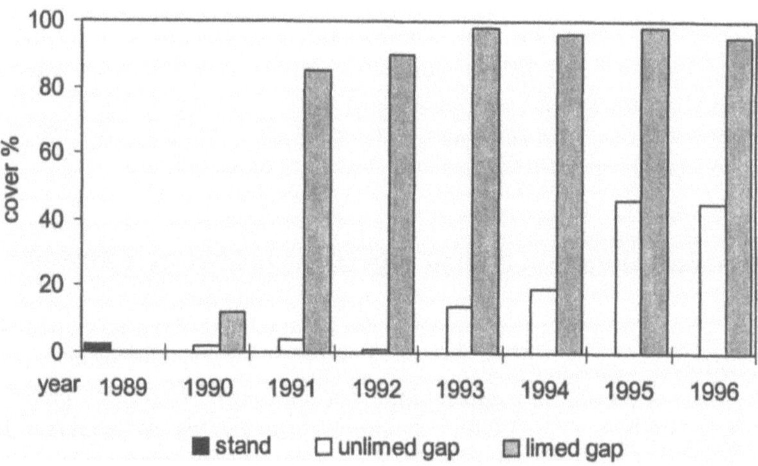

Fig. 3.1-6. Ground vegetation cover (%) in the stand, in the centre of the unlimed gap and in the centre of the limed gap.

The situation in limed gaps was totally different. A vigorous growth of herb vegetation, dominated by fireweed (*Epilobium angustifolium*) became established in the year following cutting. Other nitrophile species, which only grew in the limed gaps, are *Taraxacum officinale, Scrophularia nodosa, Urtica dioica, Epilobium montanum* and *Stellaria media*. In the second year, the percentage vegetation cover exceeded 80% in the limed gaps. In the absence of liming the vegetation cover was only about 20% in the fifth year and about 50% in the sixth year following cutting.

At the end of the 7th vegetation period after gap creation (1996), the limed gaps underwent a transitional phase from the stage of the initial population by nitrophile vegetation typical of clear-cut areas, to a stage with pioneer tree growth. While there was a marked decrease of *Epilobium angustifolium* and *Epilobium montanum*, the pioneer trees and shrubs *Salix caprea, Sorbus aucuparia, Populus tremula, Betula pendula, Rubus fruticosus, R. idaeus* and the herbaceous species *Urtica dioica* increase. In 1996, the pioneer trees formed a light, evenly spread, 4 m high cover dominated by *Salix* in the limed gaps. *Salix* did not grow in the unlimed gaps. The only pioneer trees found in the unlimed gaps were a few individuals of *Sorbus* and *Betula*. In 1996, the *Rubus*-species in the centre of the limed gaps reached cover proportions of about 30% in comparison to about 5% coverage rates in the unlimed gaps.

In the summer of the 2nd year after gap creation, the ground vegetation in the centre of the limed gaps reached an aboveground biomass of about 250 g m^{-2}. This corresponds to the biomass production determined in most of the investigations carried out 2 to 3 years after large scale clear cut (Mark and Bormann 1972, Swank 1986), after large scale windthrow (Mellert et al. 1998) and in gaps in mountainous mixed forests (Mosandl 1984). During the subsequent 5 years, the aboveground biomass of herbaceous vegetation in the limed gaps is almost the same, whereas the root biomass increases markedly (Table 3.1-2).

The proportion of *Epilobium angustiflium* in the aboveground biomass (ground vegetation and tree regeneration) in the limed gap centre decreased from 87% to 56% between 1991 and 1996. Biomass production of herbaceous vegetation was strongly reduced under the edges of the stand. This was particularly true for the northern edge with its unfavourable water budget (soil water tension in Fig. 3.1-4).

Table 3.1-2. Biomass of the ground vegetation (above- and below-ground of the herbaceous vegetation and tree regeneration for the years 1993 and 1996. Mean (arithmetic average of 10 sample plots) and standard deviation.

	Above-ground biomass (kg ha^{-1})				Below-ground biomass (kg ha^{-1})			
	HV		TR		HV		TR	
1993								
Unlimed gap								
Southern edge	0.9	±2.8	22.2	±52.2	3.9	±4.4	17.9	±41.1
Gap centre	254.7	±436.8	130.7	±174.5	89.4	±134.0	124.0	±175.8
Northern edge	0.0		8.2	±13.9	8.9	±13.2	6.4	±10.7
Limed gap								
Southern edge	1259.8	±384.0	262.2	±209.5	495.5	±379.3	264.5	±276.4
Gap centre	2644.2	±530.9	296.9	±54.3	1057.3	±544.6	18.7	±28.3
Northern edge	88.4	±63.1	30.0	±41.4	656.9	±534.5	24.1	±33.9
1996								
Unlimed gap								
Southern edge	12.1	±38.2	113.1	±168.7	38.1	±52.2	103.2	±186.0
Gap centre	646.5	±622.9	237.5	±246.9	293.5	±490.8	251.6	± 257.9
Northern edge	65.3	±120.7	66.9	±66.0	17.9	±26.7	50.9	±56.1
Limed gap								
Southern edge	1092.8	±543.6	471.1	±431.4	1339.8	±1454.9	296.3	±290.5
Gap centre	2310.0	±630.2	0.0		4138.6	±2741.7	8.8	±27.7
Northern edge	406.9	±399.3	370.2	±429.9	824.6	±655.5	346.4	±396.3

HV = herbaceous vegetation
TR = tree regeneration

Herbaceous biomass was significantly higher in limed gaps than in unlimed gaps. In the centre of the limed gap the amount measured in 1993 was about 10 times, and in 1996 about 4 times the amount measured in the unlimed gap. For the root biomass the differences between the gap variants were still greater.

The lush herbaceous vegetation in the limed gap provided severe competition for beech regeneration. The beech regeneration only reached high biomass production in gap areas with a low herbaceous biomass (Table 3.1-2). In the centre of the limed gaps, the beech plants which initially grew after the mast years 1989 and 1990 disappeared almost entirely. In the unlimed gaps by contrast, the biomass was highest in the gap centre.

The biomass of the ground vegetation represents an important factor for the element retention in the gaps. In the summer (August) of the fourth year after gap creation (1993) the vegetation (above- and below-ground biomass) in the centre of the limed gap had stored 54 kg N ha^{-1} more than the vegetation in the unlimed gap (Table 3.1-3). Three years later, the difference increased to almost 100 kg N ha^{-1}.

Table 3.1-3. Element content (N, K, Ca, Mg) of the ground vegetation (above- and below-ground herbaceous vegetation and tree regeneration) for the years 1993 and 1996. Mean (arithmetic average of 10 sample plots) and standard deviation.

	N		K		Ca		Mg	
				(kg ha^{-1})				
1993								
Unlimed gap								
Southern edge	0.6	±1.2	0.5	±1.3	0.3	±0.8	0.1	±0.1
Gap centre	9.5	±12.2	5.6	±9.0	1.8	±2.5	0.6	±0.9
Northern edge	0.4	±0.5	0.2	±0.3	0.0	±0.1	0.0	±0.1
Limed gap								
Southern edge	32.7	±11.1	27.5	±10.1	11.8	±5.1	5.9	±2.2
Gap centre	63.3	±21.7	36.1	±10.6	26.9	±10.8	15.1	±4.4
Northern edge	27.6	±18.6	15.1	±11.9	14.9	±12.6	6.8	±4.3
1996								
Unlimed gap								
Southern edge	3.0	±3.5	0.9	±1.1	0.7	±0.9	0.1	±0.2
Gap centre	22.2	±15.4	8.5	±8.9	3.6	±2.5	1.0	±0.9
Northern edge	3.7	±3.5	1.2	±1.2	0.5	±0.4	0.2	±0.2
Limed gap								
Southern edge	49.6	±22.1	25.9	±10.7	15.4	±6.6	6.5	±2.3
Gap centre	118.4	±59.0	58.0	±23.8	67.6	±39.9	21.1	±9.0
Northern edge	27.8	±15.9	9.7	±6.2	11.4	±6.1	3.3	±1.6

The K-storage in the limed gap was 6 times higher than in the unlimed gap, the Ca- and Mg-storage by about 20 times each.

The bulk of the elements stored in the ground vegetation biomass had been returned to the organic layer and decomposed within a year. Decay rates of leaf litter and roots of *Epilobium angustifolium* were investigated by using the litter bag method. Weight loss of leaf litter was 92-98% after 16 month and roots 62-84% after 12 months (Vor 1999).

3.1.6 Litter fall

The accumulation of organic matter on the forest floor represents an energy and nutrient storage pool, a diverse habitat for different types of heterotrophic organisms, and a substratum for plants (Lousier and Parkinson 1976, Vogt et al. 1986). The amount of the accumulation is dependent on litter input and the rate of decomposition.

The average annual aboveground total litter fall of the years 1990-1996 amounted to 3850 kg ha^{-1} in the closed beech stand. The differences between the years are determined by the degree of fructification. In the years with little fructification the amount of litter is about 3000 kg ha^{-1} yr^{-1}, in cases of strong masts between 4000 (1990 and 1992) and 5500 kg ha^{-1} yr^{-1} (1995).

The biomass of the leaf litter for the different years varied by 20% at the most. The differences in the litter fall between the years are due to seed drop and primarily the cupules. In most years the biomass of the seeds and cupules is almost equal to the biomass of the leaf litter (Fig. 3.1-7).

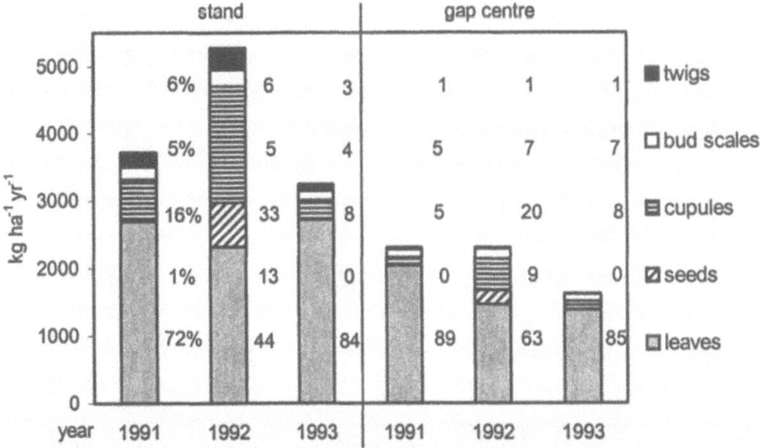

Fig. 3.1-7. Litter fractions in the stand in the centre of the gaps for the years 1991-1993. Means (kg ha^{-1} yr^{-1}) and contribution of the litter fractions to total litter fall (%).

Due to the prevailing wind direction the litter fall in the gaps decreased from the middle of the gap to the northern and eastern sides (Table 3.1-4). The low inputs of litter were mainly due to the lower inputs of heavy litter components such as seeds, cupules and twigs (Fig. 3.1-7). The easily blown litter components were relatively more represented in the middle of the gap.

Due to the amount of litter, its continual supply and the comparatively easy disintegration, leaf litter is particularly important for nutrient supply. Table 3.1-4 shows the element contents of the leaf litter of the unlimed gaps. Liming increased the Ca concentration in the leaves about 2 fold and the Mg concentration about 3 fold.

Table 3.1-4. Leaf litter input in the unlimed gap (average of 1990-1996, kg ha^{-1} and % of leaf litter in the stand, minimum and maximum in brackets) and element flux in the leaf litter (arithmetic average of the years 1990 to 1996, kg ha^{-1}).

	Leaf litter (kg ha^{-1})	Element flux (kg ha^{-1})			
		N	K	Ca	Mg
Stand	2581 (2316-2877)	28.6	13.2	10.8	1.1
Distance to gap centre					
20 m south	2501 (2301-2848) 97%	27.8	12.8	10.3	1.1
20 m north	1998 (1735-2211) 77%	21.9	9.9	8.0	0.8
20 m east	1993 (1455-2245) 77%	21.7	9.9	7.9	0.8
20 m west	2394 (2050-2732) 93%	27.3	12.5	9.8	1.0
10 m south	2058 (1482-2663) 80%	22.9	10.6	8.0	0.9
10 m north	1479 (1156-1903) 57%	17.7	7.8	6.2	0.6
10 m east	1461 (996-1923) 57%	16.2	8.0	6.3	0.6
10 m west	1996 (1537-2486) 77%	22.0	10.0	8.2	0.8
0 m	1534 (998-2188) 59%	16.8	7.7	6.1	0.6

The litter from the ground vegetation in the limed gap within 2 years compensates for the decreased litter production in the gaps compared to the closed beech stand (see Table 3.1-3). The litter of the ground vegetation is in addition of a much higher quality in terms of disintegration (narrower C/N ratio, lower lignin content, high nutrient concentration).

3.1.7 Soil fauna

In the 3rd year after cutting (1992), Lumbricidae and other saprophages such as Sciaridae, Scatopsidae and Tipulidae were most abundant in the limed gap and in the limed stand. These groups of the decomposer community contribute significantly to leaf-litter breakdown by their feeding activities (Theenhaus and Schaefer 1995). In absence of liming the average density of Lumbricidae was 10 individuals m^{-2}. Liming resulted in an increase of earthworm population to an average of 307 individuals m^{-2} (Fig. 3.1-8). It is assumed that the lower acidity of limed soil resulted in the increase of Lumbricidae. A positive correlation between the density of the Lumbricidae and the lime content in the soil was also observed in other investigations (Ellenberg et al 1986, Ammer 1992). The increase in the limed areas could be attributed to the high biomass of herbs (especially *Epilobium angustifolium*), which served as a food source. Gap creation without lime treatment did not influence composition or density of the decomposer community (Theenhaus and Schaefer 1995).

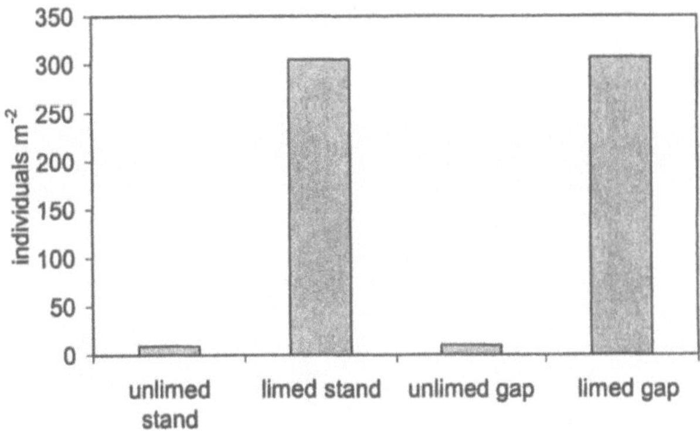

Fig. 3.1-8. Average density (individuals m^{-2}) of Lumbricidae in the centre of the unlimed gap, in the centre of the limed gap and in the surrounding unlimed and limed stand.

3.1.8 Mineralization

Because of the increased density of the decomposer community in the limed gaps, a more rapid mineralization and immobilization of nutrients was expected. This however, was not found. Also, gap creation did not increase N mineralization as repeatedly reported for large-scale clear cuts (Glavac and Koenies 1978, Vitousek and Matson 1985, Rapp 1990, Raison et al 1993). N mineralization was highest and most constant in the stand (Fig. 3.1-9).

The cumulative curves for the stand show an almost steady increase for the monitoring periods March 1991 to May 1992 and May 1996 to May 1997. In the first period a total of 90 kg N ha^{-1} was mineralized in the stand (up to 20 cm soil depth) (Bauhus and Barthel 1995). In the second period the value was 95 kg N ha^{-1} (up to 30 cm soil depth) (Vor 1999).

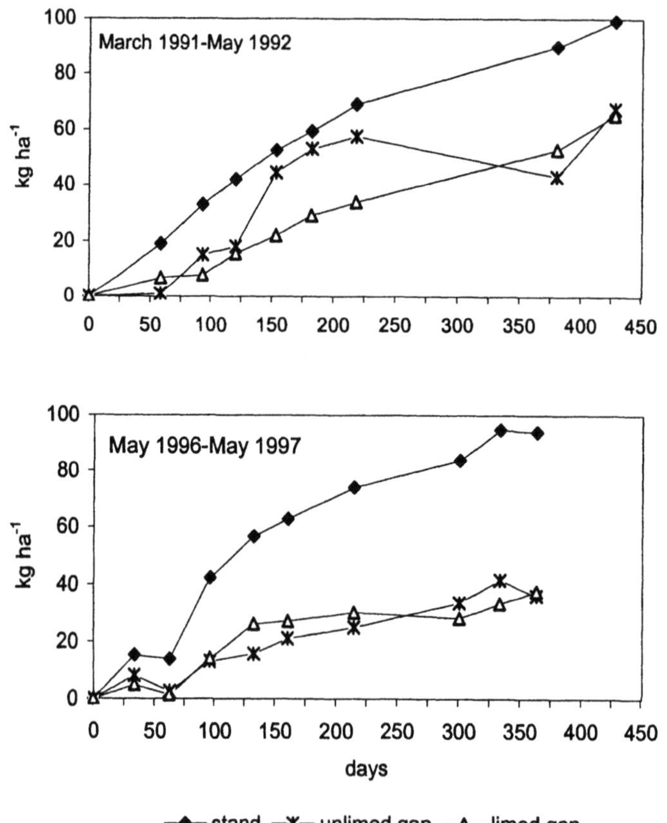

Fig. 3.1-9. Cumulative net N mineralization (kg ha^{-1}) during March 1991 to May 1992 and May 1996 to May 1997 in the stand, in the centre of the unlimed gap and in the centre of the limed gap.

About 60% of N was mineralized from the organic horizons. For the same periods N mineralization in limed and unlimed gaps was 53 and 44 kg N ha^{-1} (March 1991 to May 1992) and 36 and 38 kg N ha^{-1} (May 1996 to May 1997).

N mineralization in the gaps was not constant. Periods without net mineralization or negative net mineralization occurred. This phenomenon and the reduction in N mineralization in the gaps compared to the stand were not caused by the immobilization of N in soil microbial biomass. Liming increased microbial biomass in the beech stand, but had no significant influence on the size of the microbial pool in the gaps. Only in the seventh year after cutting, the mean N content in microbial biomass was about 15 kg N ha^{-1} higher in the limed gap than in the unlimed gap. Unlike mineralization, nitrification increased considerably as a result of tree removal and remarkably after liming. The proportion of NO_3-N in N-mineralization was 69% (1991/92) and 83% (1996/97) in the limed gap and 41% and 62% in the unlimed gap respectively. In the stand the nitrification levels reached only 20% (1991/92) and 49% (1996/97) (Bauhus and Barthel 1995, Vor 1999). The reduction of N mineralization in the gaps can be partly explained by gaseous N losses from denitrification or nitrification. Denitrification is strengthened by the high moisture content and NO_3-N concentrations (see below) in gaps and is the main source of nitrous oxide (N_2O) emissions from the acid forest soil (Bauhus et al. 1996, Vor and Brumme 2002). The N_2O emissions in the mature beech stand and in the centre of the unlimed gap ranged between 1 to 3 kg N_2O-N ha^{-1} yr^{-1}, and between 6 and 11 kg N_2O-N ha^{-1} yr^{-1} respectively. Liming reduced the N_2O losses by more than half, but increased the proportion of gaseous NO and N_2 emissions (Brumme 1995, Vor 1999).

3.1.9 Element fluxes in precipitation

Less than 80% of the precipitation above the forest canopy reached the ground of the beech stand, while about 90% reached the ground of the gap centre (Table 3.1-5). In years with substantial snowfall, such as 1991, the water input was even higher in the gap centre than in open field. As the precipitation passed through the forest canopy its chemistry changed considerably. In comparison with the element flux of open field precipitation, the element flux from canopy throughfall and stem flow (about 15% of the throughfall runs down the stems of the beech trees) was greatly enriched in potassium (K, 11 times), magnesium (Mg), sulphate (SO_4-S), hydrogen ions (H$^+$) and chloride (Cl) (about 2 times each).

Table 3.1-5. Mean and min/max of annual means for precipitation (H_2O) (L m^{-2} yr^{-1}) and element fluxes (kg a^{-1} yr^{-1}) in open field, stand (throughfall and stem flow) and centre of the gap, 1991-1996.

Flux	Open field	Stand	Gap
H_2O	1148	891	997
	837-1386	694-1022	825-1126
K	2.1	23.5	12.6
	1.1-3.2	13.6-39.3	8.9-19.5
Mg	1.1	2.3	2.2
	0.5-1.5	1.6-2.8	1.3-2.8
Ca	3.8	8.3	6.9
	2.0-5.8	5.8-11.7	5.4-7.9
H	0.14	0.22	0.17
	0.05-0.26	0.13-0.36	0.09-0.27
NH_4	8.8	10.6	10.4
	7.2-10.8	8.2-12.2	7.9-12.4
NO_3	7.3	9.8	7.9
	6.1-8.0	9.3-10.7	6.9-9.3
SO_4	10.6	18.9	14.4
	7.8-13.3	16.3-20.6	11.8-17.8
Cl	12.4	21.1	15.5
	8.1-16.4	16.9-25.6	11.5-19.5

Liming did not influence the element concentration in the canopy throughfall. Element fluxes showed considerable annual variability during the monitoring period. The increase in N inputs was comparatively low. However, it was likely that the amount of N in throughfall was an underestimation of the actual input since the gaseous component of dry deposition may be assimilated in the canopy. The average total annual N input in the beech stand in the Solling has been estimated to be 40 kg N ha^{-1} (Matzner 1989).

Although more precipitation reached the ground in the gap centre than in the beech stand, the mass input of chemicals was lower in the centre of the gap due to missing interception. As the H^+ and SO_4-S inputs indicate, less acidity was deposited in gaps. The reduced input would be even greater if the gaseous S depositions were also considered, as the uptake of SO_4-S is accompanied by an equivalent acid deposition. The input of base cations also decreased in the gaps. Since base cations may become limiting for tree growth on this strongly acidified site, reduced inputs in the gaps may affect the growth of regeneration in the phase of rapid biomass accumulation.

3.1.10 Soil solution

After gap creation, the pH of the soil solution initially dropped at 10 cm soil depth in the unlimed gap and in the limed gap (Fig. 3.1-10). In the second year the soil solution pH in the limed gap clearly exceeded that in the stand and the unlimed gap. The differences were still present in the seventh year. In the unlimed gap, the pH was constantly lower than in the stand and in the limed gap. At a soil depth of 80 cm, cutting and liming have had no significant effect on soil solution pH (Bauhus 1994, Bartsch 2000).

In comparison with the stand, a dramatic change in the element cycling in soil solution occurred in the gaps. The high nitrification and the absence of plant uptake in the first year after gap creation resulted in NO_3-N concentrations up to 3000 μmol_c L^{-1} below the humus layer, in the unlimed gap as well as in the limed gap. Similarly high concentrations could also be observed to a soil depth of 15 cm.. From the second year after the start of the experiment, the NO_3-N concentrations in the limed gap at a depth of 10 cm were below the NO_3-N concentrations of the stand (Fig. 3.1-11). With the growth of herbaceous vegetation, beginning in May, the NO_3-N concentrations decreased. The highest concentration beneath the humus layer and in the upper mineral soil layers in the stand and in the unlimed gap were measured in summer.

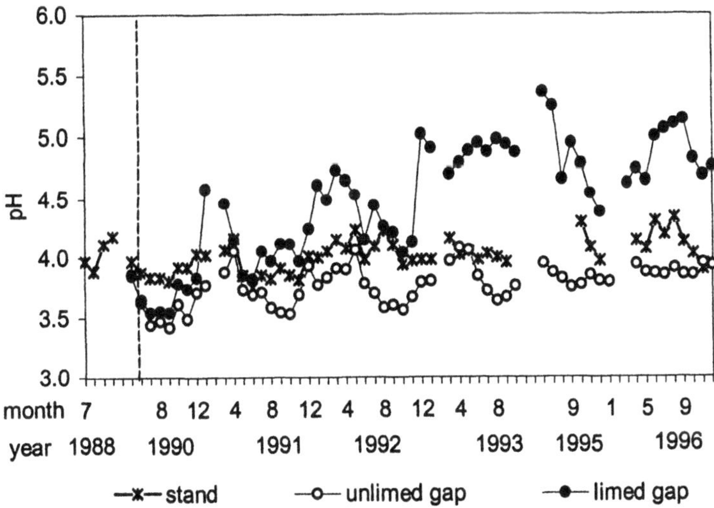

Fig. 3.1-10. pH in soil solution at 10 cm soil depth in the stand, the centre of the unlimed gap and the centre of the limed gap, 1988, 1990-1996. Dotted line indicates gap creation in October 1989.

In the mature beech stand, the NO_3-N concentrations below the rhizosphere were negligible, whereas in the gaps, high NO_3-N leaching occurred (Fig. 3.1-12). In the unlimed gap, the NO_3-N concentrations exceeded the European Union (EU) threshold value for drinking water in the first four years after gaps had been cut.

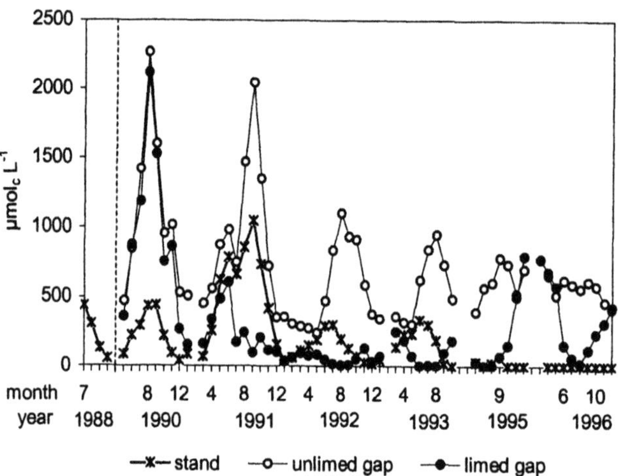

Fig. 3.1-11. NO_3-N concentrations (μmol_c L^{-1}) in soil solution at 10 cm soil depth in the stand, the centre of the unlimed gap and the centre of the limed gap, 1988, 1990-1996. Dotted line indicates gap creation in October 1989.

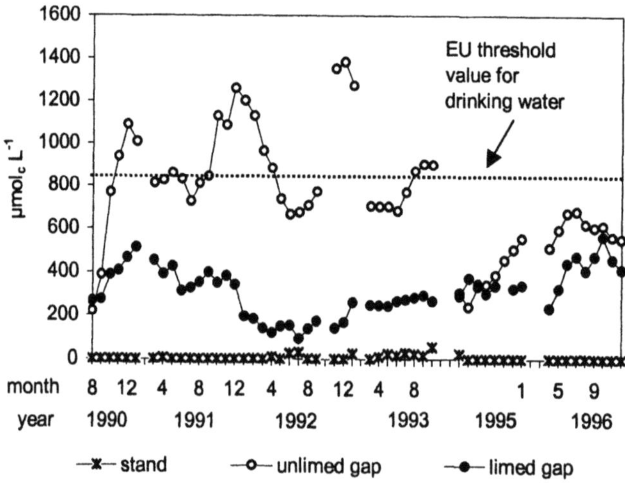

Fig. 3.1-12. NO_3-N concentrations (μmol_c L^{-1}) in soil solution below the rhizosphere (80 cm) of the stand, the centre of the unlimed gap and the centre of the limed gap. The European Union standard for maximum tolerable NO_3-N concentrations is indicated by the dotted line.

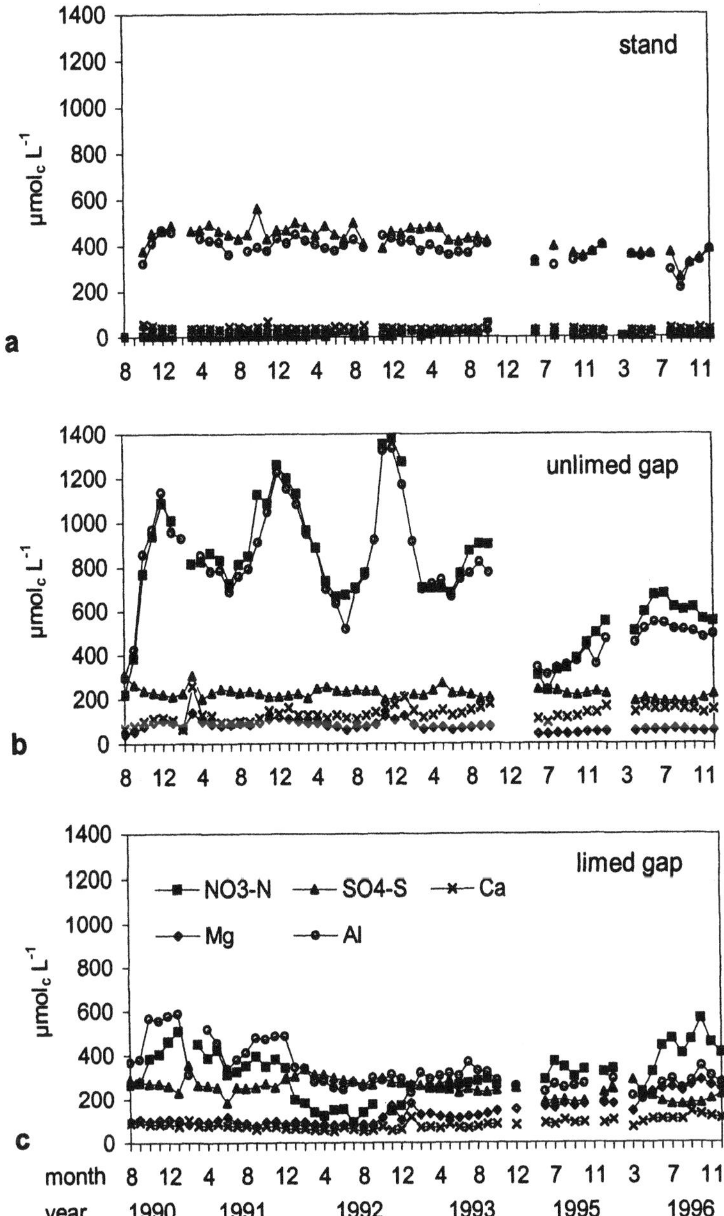

Fig. 3.1-13a-c. Concentrations (μmol_c L^{-1}) of NO_3-N, SO_4-S, Al, Ca and Mg in soil solution below the rhizosphere (80 cm) of the stand (a), the centre of the unlimed gap (b) and the centre of the limed gap (c).

Liming and associated vegetation growth reduced the NO_3-N concentrations by more than half. In the sixth year after the gaps were cut, the NO_3-N concentrations in the unlimed gap had decreased to the level of the limed gap. In contrast to the gaps, the dominant anion in the seepage water output in the mature beech stand was SO_4-S.

Since the soil chemistry is characterised by the aluminum buffer range, the main accompanying cation to NO_3-N was Al, which was leached in almost equivalent amounts (Fig. 3.1-13a-c). The leaching of base cations increased in gaps. This is a serious additional loss, because the soil is historically impoverished in base cations. The loss is certainly compensated for by liming.

The mobility of heavy metals, such as Zn and Cd, increased considerably in the unlimed gap (Fig. 3.1-14a, b). The Cd concentrations at 80-cm soil depth reached the EU threshold value for drinking water in the first three years. Liming reduced the solubility of Cd and Zn. Their concentrations in the limed gap were lower than in the beech stand.

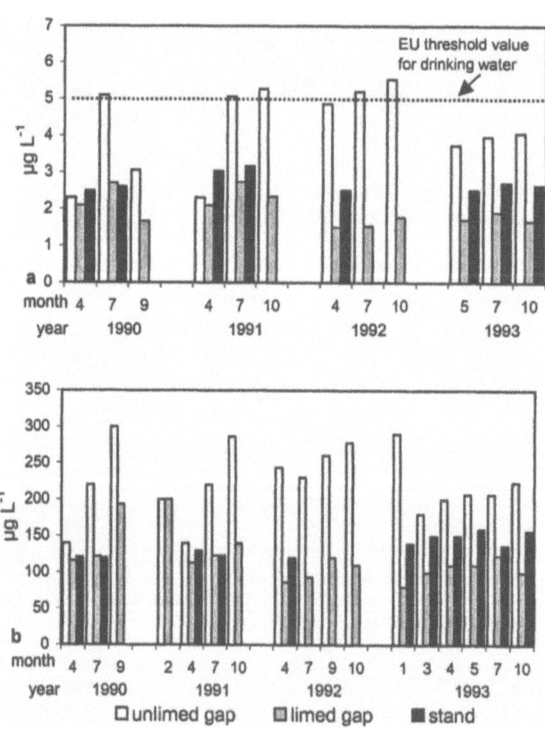

Fig. 3.1-14a, b. Cd (a) and Zn (b) concentrations ($\mu g\ L^{-1}$) in soil solution below the rhizosphere (80 cm) of the stand, in the centre of the unlimed gap and in the centre of the limed gap. The European Union standard for maximum tolerable Cd concentrations is indicated by the dotted line.

3.1.11 Input-output budgets

The increased concentrations of NO_3-N, Mg, Ca and Al in the seepage water of the unlimed gaps resulted in increased output of these elements (Table 3.1-6). Output rates NO_3-N in the unlimed gap were about 100 times higher than rates reported for the adjacent mature beech stand (Bredemeier 1987). The annual output of Al, and both Mg and Ca in the unlimed gap increased about 4 times, and 3 times over the stand, respectively. Output rates of NO_3-N in the unlimed gap were about 7 times (1992) higher than in the limed gap.

In 1996, six years after the gaps were cut, the NO_3-N output in the unlimed gap dropped by half, but remained at about 2 times greater than the annual N mineralization, which was 57 kg ha^{-1} in 1991 (Bauhus und Barthel 1995) and 36 kg ha^{-1} in 1996 (Vor 1999). While NO_3-N leaching losses in the unlimed gap decreased substantially from 1991 to 1996, NO_3-N leaching losses in the limed gap were unchanged in these years. However, in 1996 N losses were still considerably higher in the unlimed gap compared to the limed gap.

Input-output budgets of the gaps were calculated from the element flux in precipitation and the output in seepage water beneath the rhizosphere. Only K had accumulated in the gaps (Fig. 3.1-15a, b). Net losses of Mg, Ca, N, S and Al occurred each year in the unlimed gap and in the limed gap, although at a lower amount. The N losses in the unlimed gap were exceptionally high. In the limed gap the N losses were greatly reduced, approaching more or less the N input.

Table 3.1-6. Element output (kg ha^{-1} yr^{-1}) with seepage water (H_2O) (L m^{-2} yr^{-1}) in the centre of the limed gap and in the centre of the unlimed gap for 1991-1993 and 1996.

Element	Unlimed gap				Limed gap			
	1991	1992	1993	1996	1991	1992	1993	1996
H_2O	840	850	881	773	758	781	803	671
Na	23.0	12.6	10.2	7.2	11.5	8.3	8.4	11.5
K	3.5	7.2	6.5	13.2	3.8	3.1	4.5	2.0
Mg	10.4	10.1	8.6	11.1	8.8	10.1	14.1	19.8
Ca	21.2	23.9	28.7	23.9	11.3	9.5	13.6	14.9
Al	68.7	77.6	66.0	35.1	30.2	21.1	23.2	17.1
H	0.46	0.46	0.45	0.33	0.28	0.24	0.25	0.41
NH_4	0	0	0	0.27	0	0	0.04	0.22
NO_3	118.6	125.6	111.6	63.8	40.8	17.1	31.7	40.7
SO_4	30.8	30.6	32.9	25.0	31.9	37.2	33.7	22.5
Cl	36.0	24.2	16.1	10.7	19.6	10.6	14.8	12.6

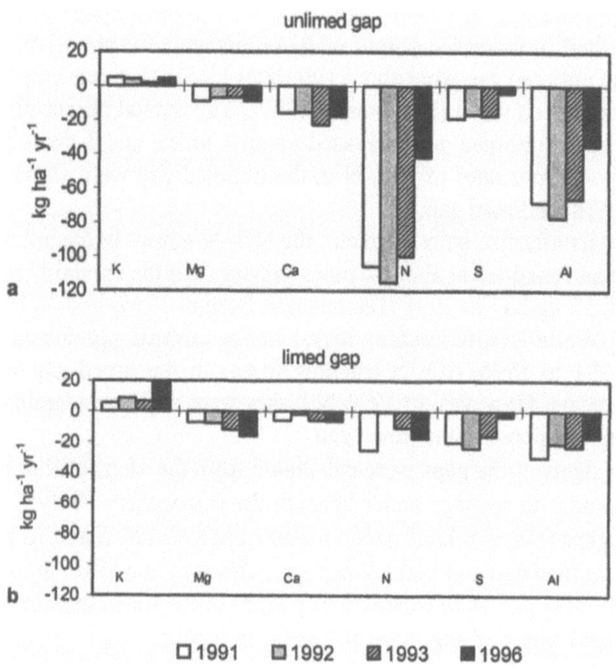

Fig. 3.1-15a, b. Elemental input-output budgets (kg ha^{-1} yr^{-1}) of the centre of the unlimed gap (a) and the centre of the limed gap (b).

3.1.12 Nitrogen standing stocks and fluxes

Nitrogen is a key element in mineral nutrition. Nitrogen is important not only in limited ecosystems, but also in N saturated ecosystems (Aber et al. 1989, 1998, Dise et al. 1998). N transformations are an important factor in the ecosystem internal proton production and consumption. Excess nitrification with subsequent nitrate leaching cause soil acidification and cation loss, while denitrification will deacidify the soil. It was shown that despite continuously high atmospheric N inputs, the mature beech stand is not N saturated, although N deposition and mineralization exceed vegetation requirements. Nitrate losses in seepage water are negligible (Meesenburg et al. 1995), and N_2O emissions are low even after N fertilisation (Brumme and Beese 1992).

The opening up of the canopy and the liming have strongly influenced the standing stocks and the fluxes of N. The most important changes in the N cycle as a result of the silvicultural treatments are shown in Table 3.1-7. Through the biomass accumulation in the gaps the ground vegetation forms an N pool, which re places the missing N pool in the beech trees in the limed gap already in the initial phase of the succession. In the limed gap a shift took place in the N storage pools from the organic layer into the mineral soil. Lower N fluxes in the gaps in comparison to the mature beech stand can be seen in the throughfall, the litter fall and mineralisation, whereas the gaseous N losses and particularly the N losses in the seepage water were increased.

Table 3.1-7. N storage pools (kg ha^{-1}) and annual N fluxes (kg ha^{-1} yr^{-1}) in the closed beech stand, in the centre of the unlimed gap and in the centre of the limed gap.

| | | Stand | Gap centre | |
	Time period		Unlimed	Limed
N storage pool (kg ha^{-1})				
Trees				
Above-ground[1]	1969-1985	700		
Roots ($\leq$ 2 mm, 0-40 cm depth)	1997	49		
Tree regeneration				
Above-ground	1993, 1996		2, 2	<1, 1
Roots	1993, 1996		1, 1	<1, <1
Herbaceous vegetation				
Above-ground	1993, 1996		5, 12	43, 50
Roots	1993, 1996		2, 7	21, 91
Microbial biomass	1991, 1996	34, 55	28, 42	24, 58
Organic layer	1997	812	738	281
Mineral soil (0-40 cm depth)	1997	5218	5070	6223
Annual N flux (kg ha^{-1} yr^{-1})				
Precipitation	1990-1996	13-19	13-19	13-19
Troughfall	1990-1996	18-23	15-22	15-22
Above-ground tree increment[1]	1969-1985	12		
Leaf litter fall	1990-1996	25-32	11-24	11-24
Mineralization	1991, 1996	90, 94	57, 36	53, 38
Denitrification	1997	3	11	5
Losses from mineral soil	1991, 1996		118, 64	41, 41

[1]from Matzner (1988)

3.1.13 Conclusions

The gap experiment demonstrated that even small-scale disturbances in the canopy of mature stands can cause substantial changes in element fluxes and may lead to drastically element losses. Because the gaps represented only a small part of a watershed and the changes occured in a limited space of time only, the overall effect on the quality of ground and spring water is not considerable. In the evaluation of this regeneration method it must be taken into account that the regeneration of the mature beech stand takes place over a long period of time (more than 40 years) during which small areas of the stand are cut.

The results also stress that the beech ecosystems on acid soils, which have been exposed to high levels of nitrogen and sulfur deposition over a long time, in this case for more than 30 years (Matzner 1989), have a low resilience with regard to nutrient retention. Flux rates of N and S in open field precipitation and total deposition in the mature beech stand at the Solling site are higher than in other European forests (Müller-Edzards et al. 1997). Between 1981 and 1994, S fluxes decreased by 1.1 kg ha^{-1} and by 3.8 kg ha^{-1} per year in open field precipitation and under spruce, respectively. Thus, over 14 years, a reduction in S deposition of about 60% has occured. Hydrogen deposition has also decreased as a consequence of the S decline. Total N deposition increased at the Solling site from 1970 to 1980 (Matzner 1989) and has decreased only slightly since then (Meesenburg et al. 1995). The reduction in S deposition has reduced the output of S in seepage water in the long term. Although the deposition of N far exceeded the demand for wood increment, about 7-14 kg ha^{-1} yr^{-1} for more than two decades, nitrate concentrations in soil solution in the Solling beech stand did not increase.

If N depositions were to continue at high levels, beech ecosystems may move towards N saturation. Significant N losses in these ecosystems may not only occur as a response to disturbance, but also as a result of N saturation in mature forest stands. This can be observed in an adjacent spruce stand and in a beech stand on basaltic bedrock (Eichhorn 1995). In combination with simultaneous losses of base cations, the forest ecosystems concerned enter phases with increasingly unstable nutrient supplies.

The extremely high N losses in the unlimed gaps correspond to the N losses in large-scale clear cuts and devegetated sites of the Hubbard Brook study (Bormann and Likens 1979). In both experiments the removal of nutrient uptake by plants played a major role in changing the soil and seepage water chemistry. In the limed gaps, the lack of nutrient uptake by trees is largely compensated for by the growth of ground vegetation and tree regeneration.

The differences in N losses could not be explained entirely by nutrient accumulation in the vegetation. The soil microbial biomass is also a potentially important pool in the N-cycle affecting N retention and release patterns. Gap creation did not increase N mineralization as repeatedly reported for large-scale clear cuts (Glavac and Koenies 1978, Vitousek and Matson 1985, Rapp 1990, Raison et al. 1993) or as a result of severe damage by windthrow (Mellert et al. 1998). Although decomposition and mineralization at this highly acidic site was not stimulated after cutting, the lack of nutrient uptake by plants, the high nitrification and the high soil water content caused substantial solute and gaseous element losses from the gaps.

With respect to silvicultural practice, the results emphasise the importance of effective and early coupling of decomposition and nutrient uptake by new vegetation after tree removal. In this study a cover of about 50% herbaceous vegetation proved to be sufficient to reduce element losses significantly. If soil acidity or gap size inhibit rapid revegetation or an expansion of the rooting system of surrounding trees into the gaps, liming may be considered as a vegetation management practice. Liming increased the resilience of the system with regard to the disruption of element cycling. Liming promoted the establishment of herbaceous vegetation and thus the nutrient retention through plant biomass production. The results however also show that the ground vegetation is a strong competitor for beech regeneration, which at may at least delay if not prevent the establishment of a new tree generation. Gaps which are so large as to prevent canopy closure within a few years should therefore only be cut if light demanding mixed forest species are to be encouraged. For example, under site conditions of the case study, above all *Picea abies*, or an advanced regeneration of shade tolerant tree species should already be present.

The gap experiment gave no indication of the mosaic cycles postulated by Remmert (1991) for central European beech woods, of a longer grass-herbaceous phase and a change in the tree species to pioneer species. Such changes in structure were only induced by liming. The vegetation development in the unlimed gaps supports assumptions from investigations of natural forests, that after a degeneration of the old trees, regeneration is from shade tolerant tree species, without the establishment of a community composed of pioneer tree species. Depending upon the site conditions, the regeneration is often dominated by beech, as in this case study. The fenced experimental areas prevented access of grazing animals, so that no statements can be made about the effects of large grazing animals on the vegetation development.

Acknowledgements

Among the many people who contributed to research at the Solling gap experiment and deserve acknowledgement, we want to give our special thanks to the following individuals: J. Schmidt, SILVAQ GmbH Marth, for modeling the water discharge, K. J. Meiwes for the provision of the deposition data from the mature beech stand, O. Godbold and H. Desmond for improving the English and the technicians M. Günther, S. Jahn, A. Nannen, K.-H. Obal, A. Södje, Ch. Suner and U. Westphal for their collaboration. The study received financial support from the German Federal Ministry of Research and Technology (BMBF) and the State of Lower Saxony.

References

Aber JD, Nadelhoffer KJ, Stendler P, Melillo JM (1989) Nitrogen saturation in northern forest ecosystems. Bioscience 39:378-386

Aber J, McDowell W, Nadelhoffer K, Magill A, Berntson G, Kamakea M, MacNulty S, Currie W, Rustad L, Fernandez I (1998) Nitrogen saturation in temperate forest ecosystems. Bioscience 48:921-934

Ammer S (1992) Auswirkungen experimenteller Beregnung und Kalkung auf die Lumbricidenfauna und deren Leistungen (Höglwaldexperiment). Forstl Forschungsber München Bd 123

Anderson MC (1964) Studies of the woodland light climate. J Ecol 52:27-41

Attiwill PM (1994) The disturbance of forest ecosystems: the ecological basis for conservative management. For Ecol Manage 63:247-300

Ban Y, Huacheng X, Bergeron Y, Kneeshaw, DD (1998) Gap regneration of shade-intolerant *Larix gmelinii* in old-growth boreal forests of northeastern China. J Veg Sci 9:529-536

Bartsch N (2000) Element release in beech (*Fagus sylvatica* L.) forest gaps. Water Air Soil Pollution 122:3-16

Barden LS (1981) Forest development in canopy gaps of a diverse hardwood forest of the southern Appalachian mountains. Oikos 37:205-209

Bauhus J (1994) Stoffumsätze in Lochhieben. Ber Forschungszentrum Waldökosysteme, Göttingen, Reihe A, Bd 113

Bauhus J (1996) C and N mineralization in an acid forest soil along a gap-stand gradient. Soil Biol Biochem 28:923-932

Bauhus J, Barthel R (1995) Mechanisms for carbon and nutrient release and retention in beech forest gaps II. The role of soil microbial biomass. Plant and Soil 168-169:585-592

Bauhus J, Bartsch N (1995) Mechanisms for carbon and nutrient release and retention in beech forest gaps I. Microclimate, water balance and seepage water chemistry. Plant and Soil 168-169:579-584

Bauhus J, Bartsch N (1996) Fine root growth in beech (Fagus sylvatica) forest gaps. Can J For Res 26:2153-2159

Bauhus J, Meyer AC, Brumme R (1996) Effect of the inhibitors nitrapyrin and sodium chlorate on nitrification and N_2O formation in an acid forest soil. Biol Fertil Soils 22:318-325

Bazzaz FA, Wayne PM (1994) Coping with environmental heterogeneity: the physiological ecology of tree seedling regeneration across the gap-understory continuum. In: Chaldwell MM, Pearcy RW (eds) Expoitation of environmental heterogeneity by plants. Academic Press, San Diego, pp 349-390

Bengtsson J, Nilsson SG, Franc A, Menozzi P (2000) Biodiversity, disturbances, ecosystem function and management of European forests. For Ecol Manage 132:39-50

Bormann FH, Likens GE (1979) Pattern and process in a forested ecosystem. Springer, New York

Botkin DB (1993) Forest dynamics – an ecological model. Oxford University Press, Oxford

Bredemeier M (1987) Stoffbilanzen, interne Protonenproduktion und Gesamtsäurebelastung des Bodens in verschiedenen Waldökosystemen Norddeutschlands. Ber Forschungszentrum Waldökosysteme, Göttingen, Reihe A, Bd 33

Brumme R (1995) Mechanisms for carbon and nutrient release and retention in beech forest gaps III. Environmental regulation of soil respiration and nitrous oxide emissions along a microclimatic gradient. Plant and Soil 168-169:593-600

Brumme R, Beese F (1992) Effects of liming and nitrogen fertilization on emissions of CO_2 and N_2O from a temperate forest. J Geophys Res 97:851-858

Büttner G (1997) Ergebnisse der bundesweiten Bodenzustandserhebung im Wald (BZE) in Niedersachsen 1990-1991. Schr Forstl Fak Univ Göttingen Bd 122.

Canham CD, Denslow JS, Platt WJ, Runkle JR, White PS (1990) Light regimes beneath closed canopies and tree-fall gaps in temperate and tropical forests. Can J For Res 20:620-631

Collins BS, Pickett STA (1988) Demographic responses of herb layer species to experimental canopy gaps on northern hardwood forest. J Ecol 76:437-450

Denslow JS, Spies T (1990) Canopy gaps in forest ecosystems: an introduction. Can J For Res 20:619

Diaci J, Boncina A (1997) Research on gap dynamics and regeneration of temperate virgin forests. Manuscript COST E4 action: forest reserves research network, pp1-14

Dise NB, Matzner E, Gundersen P (1998) Synthesis of nitrogen pools and fluxes from European forest ecosystems. Water Air Soil Pollution 105:143-154

Eichhorn J. (1995) Stickstoffsättigung und ihre Auswirkungen auf das Buchenwaldökosystem der Fallstudie Zierenberg. Ber Forschungszentrum Waldökosysteme, Göttingen, Reihe A, Bd 124

Ellenberg H, Mayer R, Schauermann J (1986) Ökosystemforschung – Ergebnisse des Solling-Projekts. Ulmer, Stuttgart

Geiger R, Aron RH, Todhunter P (1995) The climate near the ground. Vieweg, Braunschweig

Gehrmann J (1984) Einfluß von Bodenversauerung und Kalkung auf die Entwicklung von Buchennaturverjüngung (*Fagus sylvatica*) im Wald. Ber Forschungszentrum Waldökosysteme, Göttingen, Reihe A, Bd 2

Glavac V, Koenies H (1978) Vergleich der N-Nettomineralisation in einem Sauerhumus-Buchenwald (Luzulo-Fagetum) und einem benachbarten Fichtenforst am gleichen Standort vor und nach dem Kahlschlag. Oecol Plant 13:207-218

Hibbs DE (1982) Gap dynamics in a hemlock-hardwood forest. Can J For Res 12:522-527

Holeksa J (1993) Gap size differentiation and the area of forest reserve. In: Broekmeyer MEA, Vos W, Koop, H (eds) European forest reserves. Pudoc, Wageningen, pp 159-165

Knapp HD, Jeschke L (1991) Naturwaldreservate und Naturwaldforschung in den ostdeutschen Bundesländern. Schr-R Vegetationskde 21:21-54

König N, Fortmann H (1996) Probenvorbereitungs-, Untersuchungs- und Elementbestimmungsmethoden des Umweltanalytik-Labors der Niedersächsischen Forstlichen Versuchsanstalt und des Zentrallabors II des Forschungszentrums Waldökosysteme. Ber Forschungszentrum Waldökosysteme, Göttingen, Reihe B, Bd 46 and 47

Korpel S (1995) Die Urwälder der Westkarparten. Gustav Fischer, Stuttgart

Kuuluvainen T (1994) Gap disturbance, ground microtopography, and the regeneration dynamics of boreal coniferous forests in Finland: A review. Ann Zool Fennici 31:35-51

Loftfield N, Brumme R, Beese F (1992) Automated monitoring of nitrous oxide and carbon dioxide flux from forest soil. Soil Sci Soc Am J 56:1147-1150

Lorimer CG (1989) Relative effects of small and large disturbances on temperate hardwood forest structure. Ecology 70:565-567

Lousier JD, Parkinson D (1976) Litter decomposition in a cool temperate deciduous forest. Can J Bot 54:419-436

Luft W (1973) Waldbaulich-ökologische Untersuchungen bei der Femelschlagverjüngung im montanen Tannen-Buchenwald des westlichen Hochschwarzwaldes. Schriftenr Landesforstverw Baden-Württemberg Bd 39

Lüpke B v (1982) Versuche zur Einbringung von Lärche und Eiche in Buchenbestände. Schr Forstl Fak Univ Göttingen Bd 74

Lüpke B v (1987) Einflüsse von Altholzüberschrimung und Bodenvegetation auf das Wachstum junger Buchen und Traubeneichen. Forstarchiv 58:18-24

Lüpke B v (1998) Silvicultural methods of oak regeneration with special respect to shade tolerant mixed species. For Ecol Manage 106:19-26

Marks PL, Bormann FH (1972) Revegetation following forest clearcutting: mechanisms for return to steady-state nutrient cycling. Science 176:914-915

Matzner E (1988) Der Stoffumsatz zweier Waldökosysteme im Solling. Ber Forschungszentrum Waldökosysteme, Göttingen, Reihe A, Bd 40

Matzner E (1989) Acidic precipitation: case study Solling. In: Adriano DC, Havas M (eds) Acidic precipitation. Springer, New York, pp 39-81

McConnaughay KDM, Bazzaz FA (1987) The relationship between gap size and performance of several colonizing annuals. Ecolgy 68:411-416

Mellert K-H, Kölling C, Rehfuess KE (1988) Vegetationsentwicklung und Nitrataustrag auf 13 Sturmkahlflächen in Bayern. Forstarchiv 69:3-11

Meesenburg H, Meiwes KJ, Rademacher P (1995) Long term trends in atmospheric deposition and seepage output in northwest German forest ecosystems. Water Air and Soil Pollution 85:611-616

Minckler LS, Woerheide JD, Schlesinger RC (1973) Light, soil moisture and tree reproduction in hardwood forest openings. US Dept Agric For Serv Res Pap NC-89

Mlandenoff DJ (1987) Dynamics of nitrogen mineralization and nitrification in hemlock and hardwood treefall gaps. Ecology 68:1171-1180

Mosandl R (1984) Löcherhiebe im Bergmischwald. Forstl Forschungsber München Bd 61

Müller-Edzards C, de Vries W, Erisman JW (eds) (1997) Ten years of monitoring forest condition in Europe. EC-UN/ECE, Brussels, Geneva

Oliver CD (1981) Forest development in North America following major disturbances. Forest Ecol Manage 3:153-168

Parsons WFJ, Knight DH, Miller SL (19949 Root gap dynamics in Lodgepole pine forest: nitrogen tranformations in gaps of different size. Ecol Appl 4:354-362

Platt WJ, Strong DR (eds) (1989) Special feature – Gaps in forest ecology. Ecology 70:535

Otto HJ (1995) Die Verwirklichung des LÖWE-Regierungsprogramms. Allg Forstz 50:1028-1031

Raison JR, Jacobsen KL, Conell MJ, Khanna PK, Keith H, Smith SJ, Pietrowski P (1993) Nutrient cycling and tree nutrition. In: CSIRO (eds) Collaborative research in regrowth forests of East Gippsland between CSIRO and the Victorian Department of Conservation and Natural resources. Canberra, pp 8-89

Rapp M (1990) Nitrogen status and mineralization in natural and disturbed mediterranean forests and coppices. Plant and Soil 128:21-30

Rebertus AJ, Veblen TT (1993) Structure and tree-fall gap dynamics of old-growth Nothofagus forests in Tierra del Fuego, Argentina. J Veg Sci 4:641-654

Remmert H (1991) The mosaic-cycle concept of ecosystems – an overwiev. Ecol Stud 85:1-21

Röhrig E (1991) Vegetation structure and forest succession. In: Röhrig E, Ulrich B (eds) Temperate deciduous forests. Elsevier, Amsterdam, pp 35-49

Röhrig E, Bartsch N (1992) Waldbau auf ökologischer Grundlage Band 2, 6. Aufl. Paul Parey, Hamburg

Runkle JR (1982) Patterns of disturbance in some old-growth mesic forests of eastern North America. Ecology 63:1533-1546

Runkle JR (1985) Disturbance regimes in temperate forests. In: Pickett STA, White PS (eds) The ecology of natural disturbance and patch dynamics. Academic Press, Orlando, pp 17-33

Runkle JR (1989) Synchrony of regeneration, gaps, and latitudinal differences in tree species diversity. Ecology 70:546-547

Runkle JR (1992) Guideliness and sample protocol for sampling forest gaps. Gen Techn Rep PNW-GTR-283. U.S. Department of Agriculture, Forest Service, Pacific Northwest Research Station, Portland

Runkle JR, Stewart GH, Veblen TT (1995) Sapling diameter growth in gaps for two Nothofagus species in New Zealand. Ecology:2107-2117

Scherzinger W (1996) Naturschutz im Wald – Qualitätsziele einer dynamischen Waldentwicklung. Ulmer, Stuttgart

Schmidt JP (1991) Feldmeßsysteme zur Erfassung bodenphysikalischer und meteorologischer Parameter. Ber Forschungszentrum Waldökosysteme, Göttingen, Reihe B, Bd 24:173-184

Schmidt JP, Blendinger C, Lange H (1995) SilVlow – eine Modelldokumentation. Ber Forschungszentrum Waldökosysteme, Göttingen, Reihe B, Bd 42:71-83

Schmidt W (1998) Dynamik mitteleuropäischer Buchenwälder. Naturschutz u. Landschaftsplanung 30:242-249

Schütz JP (1998) Licht bis auf den Waldboden: waldbauliche Möglichkeiten zur Optimierung des Lichteinfalls im Walde. Schweiz Z Forstwes 149:843-864

Shugart HH, Smith TM (1996) A review of forest patch models and their application to global change research. Clim Change 34:131-154

Stickan W (1993) Der Kohlenstoffhaushalt von Buchen auf ökophysiologischer Basis – ein Schlüssel zum Verständnis der Wachstumsdynamik von Waldbeständen? Scripta Geobotanica 20:207-216

Swank WT (1986) Biological control of solute losses from forest ecosystems. In: Trudgill ST (ed) Solute processes. John Wiley, New York, pp 85-139

Tabaku V, Meyer P (1999) Lückenmuster albanischer und mitteleuropäischer Buchenwälder unterschiedlicher Nutzungsintensität. Forstarchiv 70:87-97

Theenhaus A, Schaefer M (1995) The effects of clear-cutting and liming on the soil macrofauna of a beech forest. For Ecol Manage 77:35-51

Ulrich B, Mayer R, Khanna PK (1979) Deposition von Luftverunreinigungen und ihre Auswirkungen in Waldökosystemen im Solling. Schr Forstl Fak Univ Göttingen Bd 58

Ulrich B (1993) 25 Jahre Ökosystem- und Waldschadensforschung im Solling. Forstarchiv 64:147-152

Ulrich B (1994) Nutrient and acid/base budget of central European forest ecosystems. In: Hüttermann A, Godbold DL (eds) Effects of acid rain in forest processes. Wiley, New York:1-50

Umweltbundesamt (1997) Daten zur Umwelt. E. Schmidt, Berlin

Veerhoff M, Roscher S, Brümmer GW (1996) Ausmaß und ökologische Gefahren der Versauerung von Böden unter Wald. E. Schmidt, Berlin

Vitousek PM, Matson PA (1985) Disturbance, nitrogen availability, and nitrogen losses in an intensively managed loblolly pine plantation. Ecology 66:1360-1376

Vogt KA, Grier CC, Vogt DJ (1986) Production, turnover, and nutrient dynamics of above- and belowground detritus of world forests. Adv Ecol Res 15:303-377

Vor T (1999) Stickstoffkreislauf eines Buchenaltbestandes nach Auflichtung und Kalkung. Ber Forschungszentrum Waldökosysteme, Göttingen, Reihe A, Bd 163

Vor T, Brumme R (2002) N_2O losses in underestimation of in situ determinations of net N mineralization. Soil Biol Biochem 34:541-544

Wagner S (1994) Strahlungsschätzung in Wäldern durch hemisphärische Fotos. Ber Forschungszentrum Waldökosysteme, Göttingen, Reihe A, Bd 123

Wagner S (1998) Calibration of grey values of hemispherical photographs for image analysis. Agric For Meteorol 90:103-117

Waring RH, Running SW (1998) Forest ecosystems – analysis at multiple scales 2nd ed. Academic Press, San Diego

Watt AS (1947) Pattern and process in the plant community. J Ecol 35:1-22

Yamamoto S-I (1989) Gap dynamics in climax *Fagus crenata* forests. Bot Mag Tokyo 102:93-114

Yamamoto SI (1996) Gap regeneration of major tree species in different forest types of Japan. Vegetatio 127:203-213

3.2 Canopy disintegration and effects on element budgets in a nitrogen-saturated beech stand

J. Godt

Faculty of Urban and Landscape Planning, Department Landscape Ecology and Soil Science, Kassel University, Gottschalkstr. 28, D-34109 Kassel, Germany, e-mail: jgodt@hrz.uni-kassel.de

Abstract

The results of this project describe the effects following disturbances in a N-saturated beech stand. It could be shown that canopy disintegration leads to a disruption of the element cycle, mainly controlled by N - transformation processes. Regarding soil as black box, input/output balances in this project can describe the immediate consequences of disturbance processes as related to forest dieback or silvicultural practice.

Key words: *Fagus sylvatica*, nitrogen, element budget, N-saturation, soil acidification, nitrification, drinking water quality

3.2.1 Introduction

Defoliation and formation of small leaves often are described as symptoms of forest decline, especially for deciduous trees. These symptoms occur on acidic soils as well as on soils with high base saturation, e.g. soils on basaltic bedrock. Canopy disintegration will lead to a stronger insolation and, thereby, a warming up of the forest floor as a short term effect. A consequence of this will be accelerated decomposition processes of forest litter and organic material in the upper soil horizons especially in soils with higher pH. Changes in element budgets are to be expected. Most intensive effects can be expected in those ecosystems characterised by a high turnover of organic matter in the organic layer and upper mineral horizons. Those systems often are dominated by the N-turnover. If the N-turnover is accelerated, high NO_3^- concentrations, NO_3^- leaching and losses of basic cations from soil will lead to soil acidification, especially in N-saturated sites. In the literature, N-saturation in forest ecosystems is defined in different ways based on nutritional aspects or acidity balance approaches (e.g. Nilsson 1986, Skeffington and Wilson 1988, Agren and Bosatta 1988, Cole et al. 1992, van Miegroet et al. 1992, Nilsson and Grennfelt 1988). N-saturation given by conservative definitions

is reached, when output exceeds input rates. Discussing the Critical Load Concept
for Nitrogen, where N-saturation plays a role, Grennfelt and Thörnelöf (1992)
recommend a definition, where "the availability of inorganic N is in excess of the
total combined plant and microbial nutritional demand". This definition can be
looked upon as being the most restrictive.

3.2.2 Objectives

In this project N-mobilisation is studied as an effect of canopy disintegration in an
old beech stand on basaltic bedrock (brown earth). Canopy disintegration was
simulated by felling 11 old trees, hereby forming a gap in the formerly closed canopy structure. The main goal of the project was to quantify N-, macro- and trace
element input-/output balances for the soil compartment and acidification of soil.
The results of this experiment should give a base for understanding the reaction of
N-rich forest ecosystems to disturbance, especially with regard to the nitrogen
budget.

Such disturbances may originate from silvicultural practices such as clearfelling, gap felling or thinning, or may be caused by the breakdown of forest stands,
induced by air pollution.

As nitrogen cycling is governed by a number of interrelated processes (see Fig.
3.2-1) the results of the experiment can not be expected to give definite answers
with respect to whole forest stands under various conditions. Neither will the project be able to give definite data concerning single processes within the soil. The
soil compartment must be regarded in the present study as a black box which is
investigated through input and output fluxes. But the project allows estimation of
the risk of excessive nitrate leaching and the extent of soil internal production of
acidity. In addition, the risk for groundwater contamination is evaluated. On the
other hand the results are important in the discussion of the 'critical loads' and it's
relevance for non-static environmental conditions.

3.2.3 Methods and site description

From Sept. 1989 until May 1997 fluxes of elements have been measured in two
plots, divided in different subplots of a 150 year old beech stand on a N-rich site on
basaltic bedrock in a rural region close to Kassel, Germany. Site characteristics can
be seen in Table 3.2-1. Cation exchange capacities and saturation of the exchange
sites by acidic (c_a) and basic (c_b) cations are shown in Table 3.2-2. High base saturation (nearly 100 %) was found in deeper soil horizons, whereas significant levels
of acidic cations (mostly Al, 16%) were found in the upper horizons. The forest
floor of the site shows symptoms of N-mobilisation, indicated by the dominance of
nitratophile ground vegetation such as *Sambucus racemosa* and *Urtica dioica*.

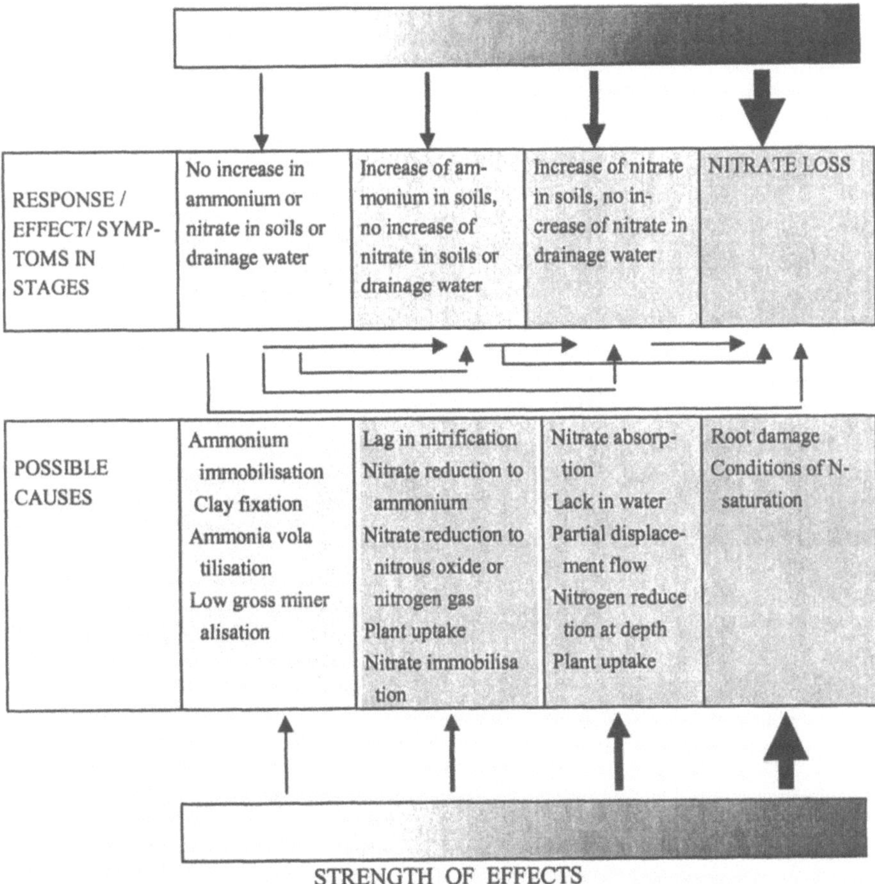

Fig. 3.2-1. Disturbances and N-cycle in forest soils (after Vitousek et al. 1982).

To collect precipitation, polyethylene bottles with a funnel fixed on top (100 cm^2 surface area) in order to protect against evaporation losses have been used in summer, open buckets in winter (see also Meiwes et al. 1984). Output is calculated from measurements in soil solution (ceramic cup suction lysimeters) at 10 cm (n = 54), 30 cm (n = 27) and below the rooting zone in 100 cm mineral soil depth (n = 27). A convenient and appropriate simulation programme for water balances in deciduous forest ecosystems taking into account both the strong inclination of the slope and the role of stemflow (see also Lischeid 1995) was not available. So calculations of the water flow in soil have been made by using Cl⁻ as an inert tracer. Elements have been measured by using IC and TOC/IC-analysers (anions), ICP (macroelements) and AAS-HGA (trace elements) and ion-sensitive electrodes (NH_4^+). Temperature measurements were done in both plots hourly, soil water potential measurements once a week, light intensity inside the stand and in an adjacent open field plot was measured once a year. Input rates are calculated according to Ulrich (1983) on an annual basis (May–Apr.) from deposition data from canopy drip (n = 9/plot), stemflow (n = 2/plot) and litterfall (n = 6/plot). For the calculation of total N deposition rates, different approaches are used in literature. In this study, deposition rates of N in canopy drip + stemflow are taken as equivalent to total deposition. Open field deposition data are taken from measurements in a neighbouring open field plot (Eichhorn and Paar 1996) in 1990-1995. In 1996, data are taken from our own measurements in district 14 in the direct neighbourhood.

Tab. 3.2-1. Site characteristics.

Location	Northern part of Hessen, Germany, close to Kassel, Community forest of Zierenberg, district 16 3518953 m right, 5690985 m high GK
Altitude above sea level	520-560 m
Inclination	17-28 degrees, NE/SE
Soil type	Deeply developed brown earth on basaltic bedrock, very rich in stones (basalt debris)
Tree species, age	Beech (*Fagus sylvatica L.*), 150-160 years old
Bonity and stems/ha	II, 123 in/around gap, 179 in reference area
Research plots	Control, gap (edge and centre)
Disturbance	Cutting of a gap in the canopy layer, 30 m in Diameter in Sept. 1990

Manipulation experiment: disturbance by cutting a gap

In order to simulate canopy disintegration, one plot has been manipulated by felling 11 trees in Sept. 1990, producing a gap of around 30 m in diameter.

Under real environmental conditions of thinning procedures or declining stands, root functioning is still existant. Therefore, the measurement plots of the gap area were placed at the edge of the gap area. Here root systems of ground vegetation and of trees surrounding the gap extending their root system into the gap area, were still active. Exposed to south, insolation in the edge of the gap under the crown cover was high. Devices for gathering stemflow and canopy drip and seepage in 10, 30 and 100 cm depth of the mineral soil were installed. Figure 3.2-2 shows the canopy structure after manipulation.

Table 3.2-2. Mineral soil characteristics (mean and standard deviation (SD) of 6 profiles).

		Depth in mineral soil					
		0-10 cm	10-20 cm	20-40 cm	40-60 cm	60-80 cm	80-100 cm
pH (NH$_4$Cl)	Mean	4.3	4.4	4.7	5.0	5.27	5.30
	SD	0.15	0.12	0.22	0.14	0.20	0.13
CEC$_e$ (μmol$_c$ g^{-1})	Mean	124.1	172.5	183.9	161.9	140.1	180.0
	SD	11.68	58.01	63.45	67.98	84.31	90.65
Ca (%)	Mean	47.6	58.2	60.6	54.7	60.0	49.6
	SD	8.59	13.86	8.31	14.64	10.59	17.04
Mg (%)	Mean	24.4	21.6	28.7	40.4	36.4	45.1
	SD	6.78	6.00	7.72	15.18	11.07	15.78
K (%)	Mean	5.7	3.4	1.9	0.8	0.6	1.3
	SD	1.49	2.22	1.79	0.6	0.32	1.16
Na (%)	Mean	1.1	1.2	1.6	2.1	2.2	3.4
	SD	0.40	0.48	0.5	0.45	0.42	1.60
H (%)	Mean	1.2	0.3	0.0	0.0	0.0	0.0
	SD	0.85	0.29	0.03	0.0	0.0	0.0
Al (%)	Mean	15.6	13.3	5.7	1.5	0.5	0.6
	SD	8.88	12.24	6.94	1.65	0.77	0.89
Fe (%)	Mean	0.8	0.1	0.1	0.0	0.00	0.0
	SD	1.15	0.12	0.16	0.06	0.01	0.01
Mn (%)	Mean	3.7	2.0	1.5	0.6	0.2	0.1
	SD	2.28	1.36	1.31	0.38	0.15	0.05
Base saturation (%)	Mean	78.7	84.4	92.6	97.9	99.2	99.3
	SD	13.01	13.72	8.25	1.98	0.78	0.91
N (%)	Mean	0.4	0.3	0.2	0.1	0.1	0.0
	SD	0.14	0.07	0.08	0.05	0.05	0.01
C (%)	Mean	6.0	5.1	3.4	2.1	1.0	0.6
	SD	2.01	0.57	1.09	0.97	0.68	0.14

3.2.4 Results

In the following figures (boxplots), maximum values are shown in the upper bar, minimum values in the lower bar, 75 percentile (top of box) and 25 percentile (bottom of box) and median in the center.

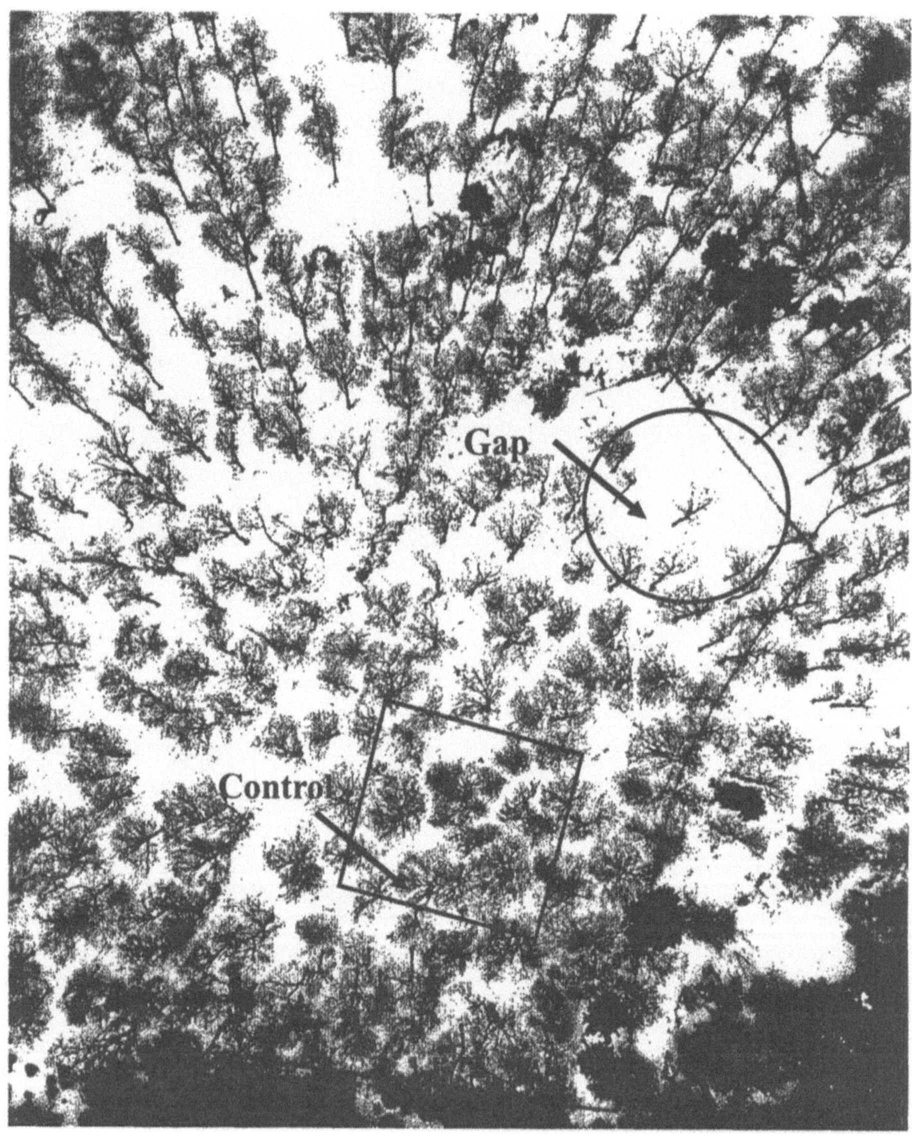

Fig. 3.2-2. Stand from the air after felling 11 trees in the gap (district 16).

Light intensity

The influence of gap in canopy structure is demonstrated by radiation measurements during the growing season (Fig.3.2-3).

One year after the manipulation, the *reference area* had an average light intensity relative to open field of 10.1%, in the gap area 20.4%, both at breast height. Differences could be shown clearly until 5 years after manipulation (data for 1996 missing). On the forest floor (measurements started in 9/92), differences between gap and control areas could only be observed for the first two years after manipula-

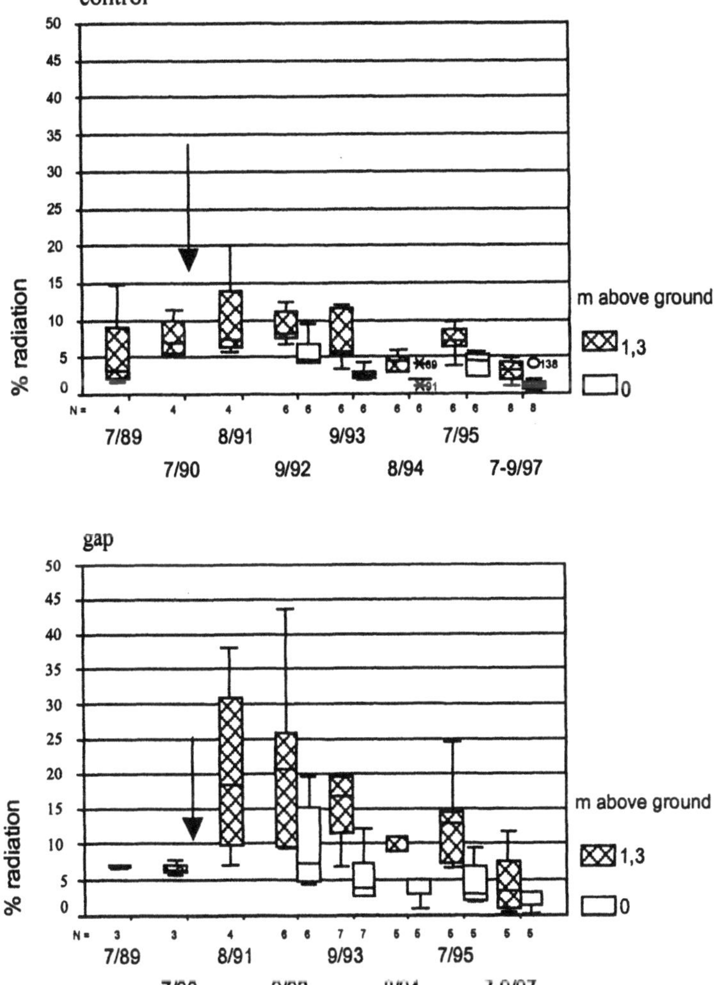

Fig. 3.2-3. Relative light intensity in gap and control areas on the floor (0 cm) and at breast height (1.3 m), ⟶ formation of gap in Sept. 1990.

tion, showing a broader variability in the gap. Maximum differences ranged in an order of about 4.1% (compared to open field = 100%) in 1992. The data directly reflect the influence of developing ground vegetation, mostly formed by *Urtica dioica, Rubus idaeus,* and *Galium odoratum* in the herb layer and mostly formed by *Sambucus racemosa* and *Fraxinus excelsior* in the understory tree layer. The understory has reached breast height about 6 years after manipulation. Already 2 years after manipulation the development of the ground vegetation is more rapid and there is a significantly reduced radiation to the ground.

Temperature

Temperature measurements have been conducted during the first 2½ years after manipulation. Figures 3.2-4 and 3.2-5 show mean temperatures for the summer and winter season in ambient air, in 5 and 30 cm mineral soil in the gap and the control area, respectively. Data from Dec. 1991 to May 1992 are missing due to a system failure. Average temperatures in ambient air did not show any clear differences while slight differences could be seen in 5 cm and 30 cm soil depth. During winter, mean daily temperatures in 5 cm mineral soil depth were colder in the gap by 1 °C, while in summer the gap area showed 3 °C warmer conditions (Fig. 3.2-6).

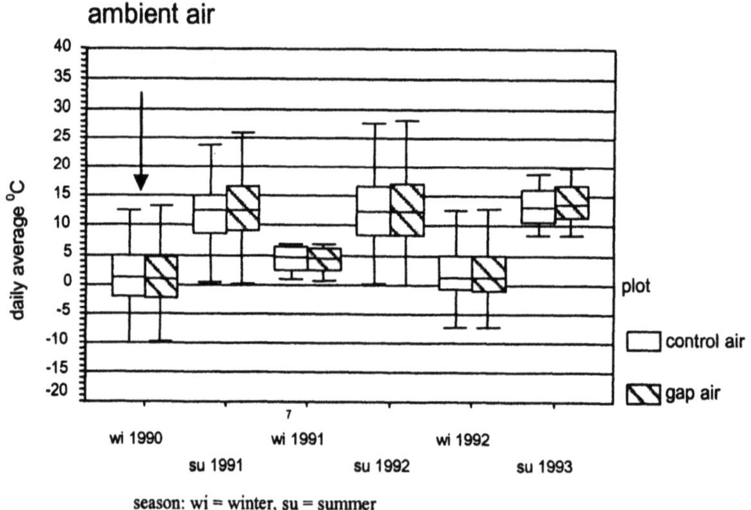

Fig. 3.2-4. Mean daily temperatures in ambient air, ⎯→ formation of gap in Sept. 1990.

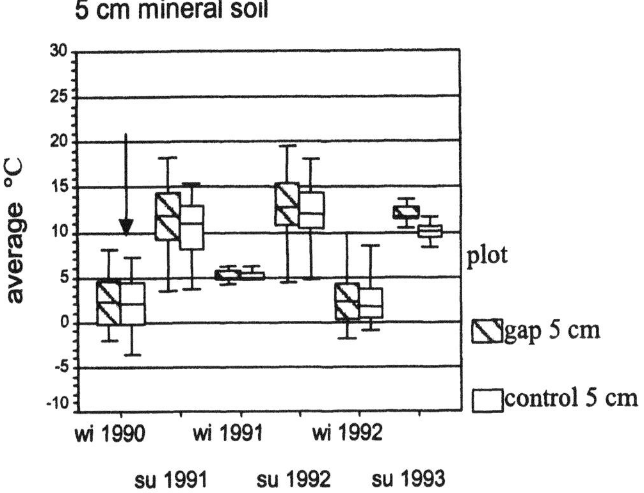

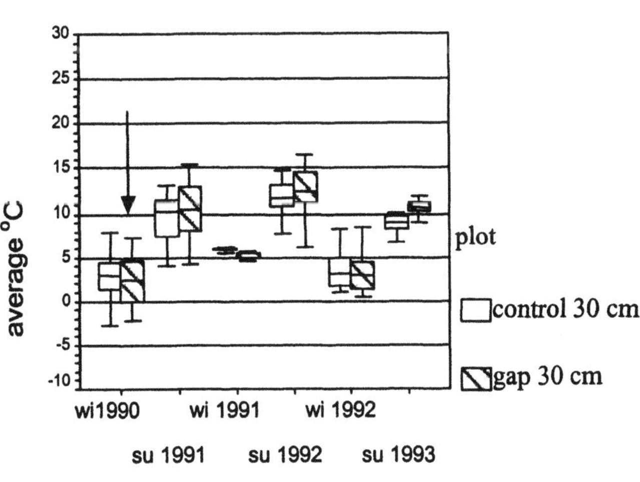

Fig. 3.2-5. Mean daily temperature in 5 cm and 30 cm soil depth of gap and reference area, ⟶ formation of gap in Sept. 1990.

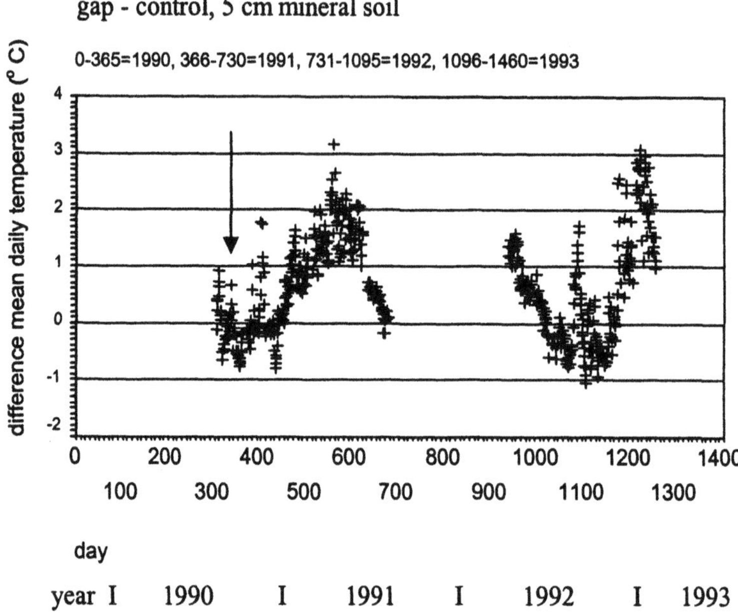

Fig. 3.2-6. Differences in mean daily temperatures in 5 cm mineral soil (gap – control), ——▶formation of gap in Sept. 1990.

Nitrogen

In Figure 3.2-7 data for NO_3^--concentrations in soil solution in 10, 30 and 100 cm mineral soil depth are given for the growing season and winter. During the first years after manipulation, higher NO_3^--concentrations in soil solution in the gap area can be seen in 100 cm only, increasing about 3 months after the disturbance up to 100 mg/L. Clear differences in NO_3^--concentrations can be seen in the period from summer 1995 until winter 1996/1997. During this time, higher NO_3^--concentrations were found in the control plot, often reaching the highest concentrations. In general, NO_3^--concentrations very often exceed the EC and national drinking water quality standard (50 mg NO_3^- /L). This is true even for the control area, especially in 100 cm depth, showing high nitrate concentration in seepage water leaving the root zone.

pH

pH values in soil solution did not show any strong acidification pushes (Fig. 3.2-8).

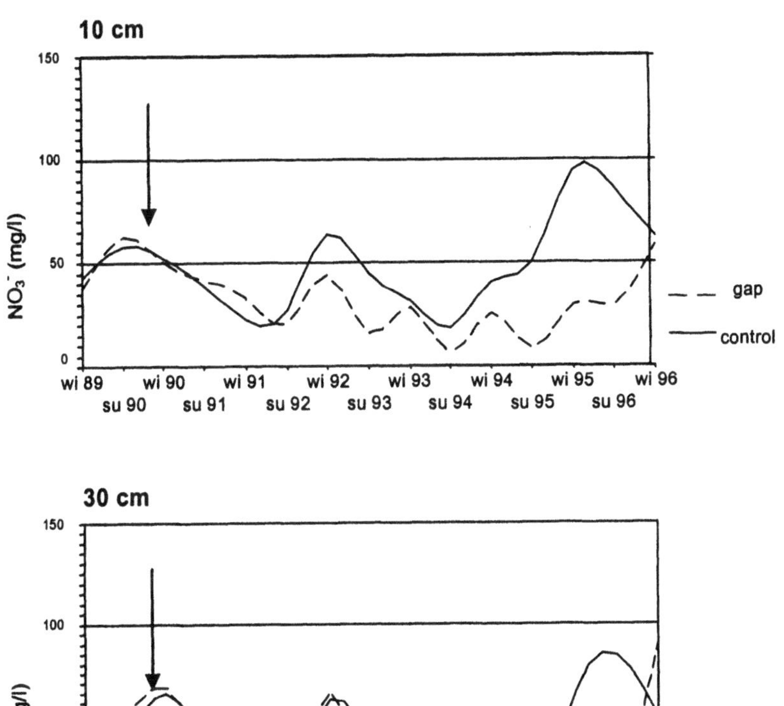

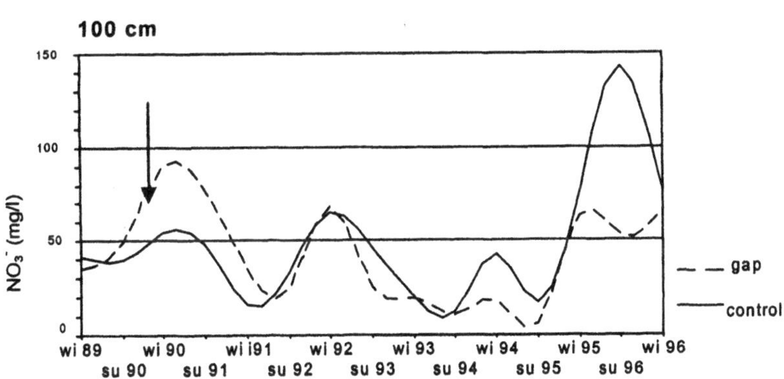

Fig. 3.2-7. NO_3^- - concentrations in soil solution in 10, 30 and 100 cm soil depth, gap and control, ⟶ formation of gap in 9/1990; wi = winter, su = summer.

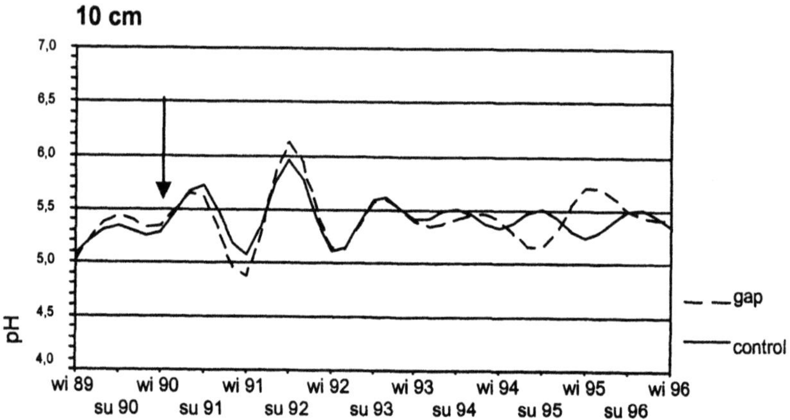

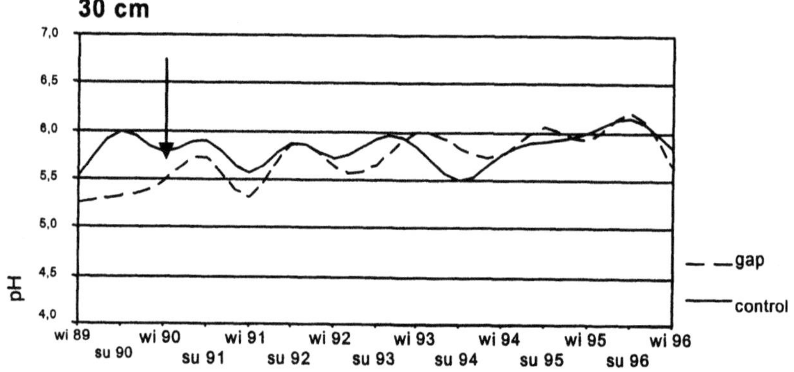

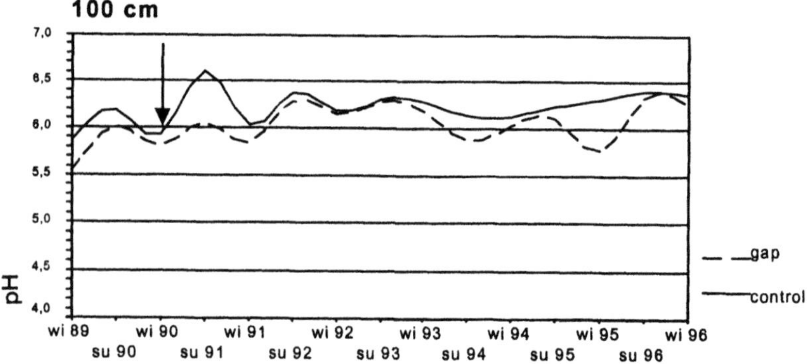

Fig. 3.2-8. Mean pH in soil solution in 10, 30 and 100 cm mineral soil depth, hydrological season: wi = winter, su = summer,⟶ formation of gap in Sept. 1990.

Water fluxes

Water fluxes in precipitation have been taken from measurements of canopy drip, stemflow and open field precipitation. Water fluxes in soil have been calculated by using Cl⁻ as an inert tracer. In these calculations, increasing Cl⁻-concentrations in seepage relative to canopy drip directly reflect the transpiration losses. Data on an annual basis can be seen from Figure 3.2-9. Output rates correspond indirectly to mm precipitation in canopy drip and stemflow.

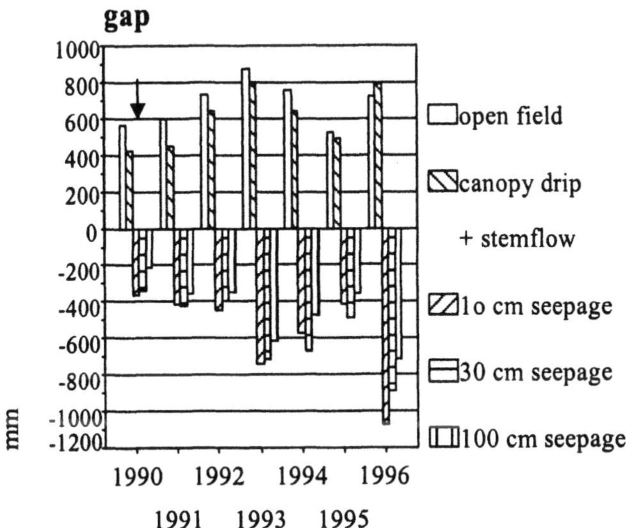

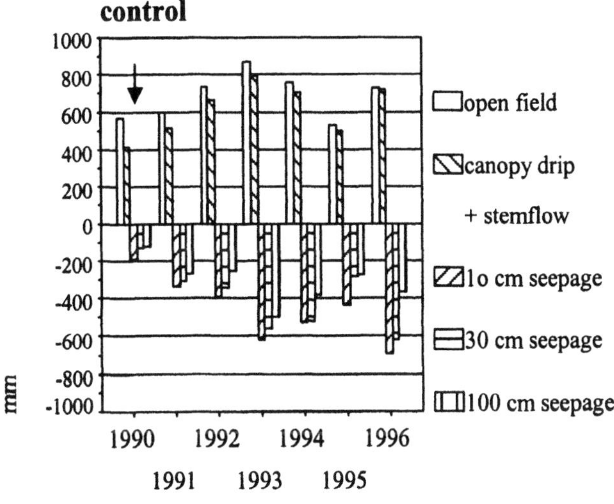

Fig. 3.2-9. Water fluxes at the edge of gap and control, ⟶ formation of gap in Sept. 1990.

Lischeid (1995) studied water balances and movement in soil in a beech stand with similar site conditions some kilometers away from the research plot (district 14). In his intensive studies he concluded from water fluxes in springs around his research area on seepage leaving root zone of 167 mm in 1991 and 175 mm in 1992. In this project water flux rates under the root zone have been calculated in the order of 267 mm (1991) and 254 mm (1992). As the site of this study is located in an altitude of around 550 m, nearly 100 m higher than in Lischeid's study, higher open field precipitation (ca. 70 mm) and less evaporation losses and thereby higher water flux rates leaving the root zone can be expected. Under these considerations the calculated water fluxes can be accepted as realistic.

N fluxes

In contrast to concentration patterns of NO_3^- in soil percolation water, N fluxes in the forest gap compared to the undisturbed plot show clear differences already in the first year, although disturbance took place in November 1990, which means that the disturbing effects could only play a role during half of the time period of the year 1990/1991 (Fig. 3.2-10, Table. 3.2-3). Comparing input (canopy drip + stemflow) and output rates (100 cm soil depth) of N-compounds it can be seen, that all of the NH_4^+-input is transformed into NO_3^-, or taken up by roots / microorganisms or stored within soil. N is leaving soil in form of NO_3^-. In the control plot the system is loosing nitrogen by leaching at a rate almost equivalent to total deposition. Calculating net balance for 1990 until 1996, the soil has lost 1.26 $kmol_c$ N ha^{-1} yr^{-1}. This fact allows the undisturbed reference site to be qualified as "nitrogen saturated", taking the definition of Agren and Bosatta (1988). More than twice the NO_3^- has left the root zone during the first 4 years after disturbance in the gap area compared to undisturbed plot. In 1996, the gap area lost the highest amount of N: 4.95 $kmol_c$ N ha^{-1} yr^{-1}. From 1990 to 1996, the system lost 7.49 $kmol_c$ N ha^{-1} yr^{-1} in net.

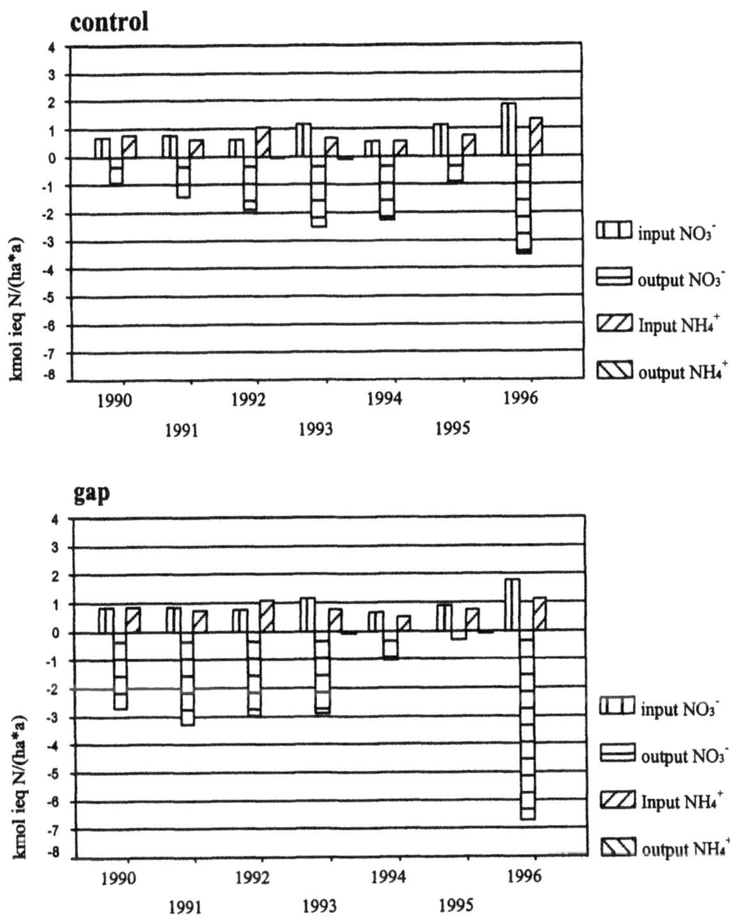

Fig. 3.2-10. N-fluxes in gap and reference area; formation of gap in Sept. 1990.

Soil balances for cations and anions

Net soil balances (input–output) can be seen from Table 3.2-3. Input rates of acidic cations (C_a: sum of H^+, NH_4^+, Mn^{2+}, Fe^{2+}, Al^{3+}) is leading to storage of acidity, accompanied by loss of base cations (C_b; sum of Ca^{2+}, Mg^{2+}, K^+, Na^+). As NH_4^+ is completely stored/transformed within the soil compartment, accelerated NO_3^--leaching is resulting. In order to maintain a charge equilibrium, high losses of C_b-cations, resulting from exchange processes, are the consequence. For SO_4^{2-} a mobilisation effect can be observed in the disturbed plot.

Table 3.2-3. Soil balances for cations and anions in control and gap.

Plot	Hydro-logical year	Acidic cations	Basic cations	NH_4^+	NO_3^-	SO_4^{2-}
				$kmol_c\ ha^{-1}\ yr^{-1}$		
Control	1990/91	+1.48	-1.13	+0.75	-0.24	+0.16
	1991/92	+1.56	-2.19	+0.58	-0.65	-0.34
	1992/93	+1.69	-2.49	+1.00	-1.25	-0.35
	1993/94	+1.24	-6.21	+0.56	-1.32	-2.02
	1994/95	+0.94	-2.88	+0.52	-1.72	-0.93
	1995/96	+1.37	-2.51	+0.71	-2.00	-0.32
	1996/97	+1.50	-7.71	+1.29	-1.65	-1.83
	Sum	+9.78	-25.12	+5.39	-6.65	-5.63
Gap	1990/91	+1.91	-3.01	+0.85	-1.85	-0.05
	1991/92	+1.59	-5.06	+0.72	-2.44	-1.12
	1992/93	+1.47	-2.73	+1.04	-2.23	-1.08
	1993/94	+1.54	-7.41	+0.67	-1.73	-2.69
	1994/95	+0.91	-2.75	+0.48	-0.36	-1.69
	1995/96	+0.68	-1.84	+0.65	+0.54	-0.94
	1996/97	+1.46	-8.54	+1.11	-4.95	-6.44
	Sum	+9.56	-31.24	+5.53	-13.02	-14.01

In Table 3.2-4 and Fig. 3.2-11 production of acidity by N-transformation and net balances for acidic cations is shown. The production of acidity by N-transformations is calculated according to Breemen et al. (1983) by calculating $((NH_4^+output - NH_4^+input) - (NO_3^-output - NO_3^-input))$. A negative value for this quantity suggests that nitrogen transformations are generating H^+ to the ecosystem (Driscoll and Schaefer, 1989). The loss of NO_3^- by leaching in the control plot results in an average (1990-1996) net acidification of about 1.98 kmol H^+ ha^{-1} yr^{-1}. Within the disturbed plot, additional acidification by N-transformation in the range of +0.66 kmol H^+ ha^{-1} yr^{-1} has been calculated.

When adding the store of acidic cations (not including NH_4^+ which has already been included in the calculation of N-transformations) a yearly mean net acidification of the soil compartment of 2.6 kmol H^+ ha^{-1} yr^{-1} for the undisturbed, and 3.2 kmol H^+ ha^{-1} yr^{-1} for the disturbed plot was estimated as an annual mean for 1990-1996. A maximum value was calculated for 1996 (6.4 kmol H^+ ha^{-1} yr^{-1}) in the gap.

Table 3.2-4. Production of acidity by N-tranformations and net balances for acidity.

Plot	Year (May-April)	Prod.of acidity by N-trans-formations	Input-output balance of acidity apart from N-compounds $(C_a\text{-}NH_4)$	Sum of acidity produced / stored in soil
			------------- $kmol_c$ ha^{-1} yr^{-1} --------------	
Control	1990/1991	0.99	0.73	1.72
	1991/1992	1.23	0.98	2.21
	1992/1993	2.26	0.69	2.95
	1993/1994	1.87	0.68	2.55
	1994/1995	2.24	0.42	2.66
	1995/1996	0.51	0.66	1.17
	1996/1997	4.74	0.21	4.95
	Sum 1990-1996	13.84	4.37	18.21
Gap	1990/1991	2.70	1.06	3.76
	1991/1992	3.15	0.87	4.02
	1992/1993	3.26	0.43	3.69
	1993/1994	2.39	0.87	3.26
	1994/1995	0.84	0.43	1.27
	1995/1996	0.11	0.03	0.14
	1996/1997	6.05	0.35	6.40
	Sum 1990-1996	18.50	4.04	22.54

Nitrogen store in soil and plant

Nitrogen stores in different soil and plant compartments at a given date (Aug. 1994) have been determined by taking soil and vegetation samples from 10 connected and representative 1 x 1 m-subplots in the centre of the gap, at the edge of the gap, and in the control, respectively. The results are expressed on a hectare basis (Table 3.2-5). The ground vegetation includes the fast developing understory, mostly formed by *Sambucus* and in some cases *Fraxinus excelsior*.

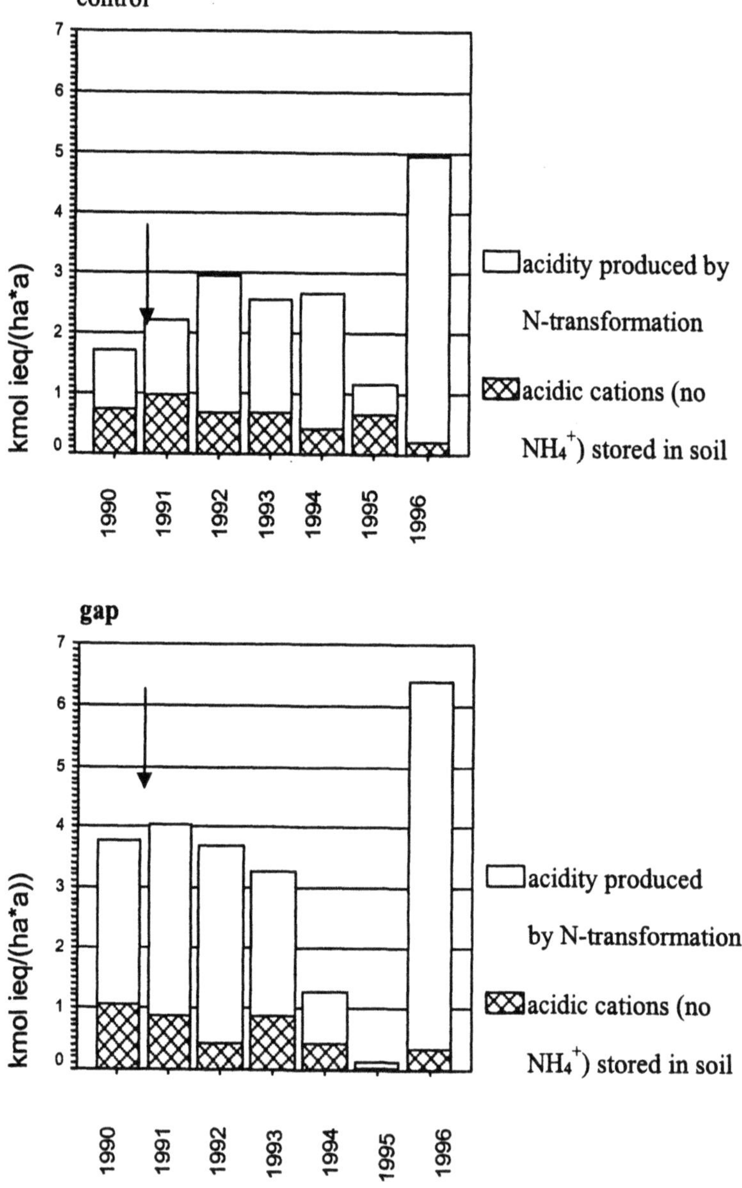

Fig. 3.2-11. Production of acidity by N-transformations in soil and storage of acidic cations in soil; ⟶ formation of gap in 9/1990.

Table 3.2-5. N-storage in different compartments in the centre and edge of gap and control area. Samples taken August 8, 1994; mean, standard deviation in brackets, and % of total sum (N-store).

	Compartment	Center of gap	Edge of gap	Control
Biomass (kg ha^{-1})	Ground vegetation incl. developing understory	11727 (23102)	3170 (2948)	473 (525)
	Organic layer	68611 (15850)	87718 (22331)	62530 (15121)
	Roots organic layer	817 (511)	777 (400)	406 (220)
	Roots mineral soil (0-25 cm)	3168 (2271)	3602 (2421)	5395 (2571)
N-store (kg ha^{-1})	Ground vegetation incl. developing understory	715.6 (1221) 17%	177 (147) 3%	17.9 (13.5) 0.4%
	Organic layer	1727 (1139) 42%	1862 776.7) 34%	768.7 (229) 15%
	Roots organic layer	15.77 (14.62) 0.4%	11.68 (6.16) 0.2%	16.04 (6.06) 0.3%
	Roots mineral soil (0-25 cm)	46.88 (26.90) 1.1%	42.10 (23.91) 0.8%	85.84 (56.01) 1.7%
	Mineral soil (0-25 cm)	1631 (643) 39%	3382 (2401) 62%	4166 (3995) 82%
	Sum	4137 100%	5475 100%	5054 100%

High standard deviations in all parameters show the methodological difficulties in calculating reliable data based on storage determinations in ecosystems, although relatively large efforts have been made to calculate these data. Tested by Scheffe-procedure, statistical differences (p $\leq$0.05) could only be found for:
- N-store in organic layer: control/edge of gap, control/center of gap
- N-store in soil: control/edge of gap, control/center of gap and edge of gap/gap
- N-store in roots in soil: control/edge of gap.

3.2.5 Discussion

It has been shown that opening of the forest canopy by cutting a gap was followed by warming the upper soil layer during summer by 1 to 2 K. Light radiation at breast height and on the ground is directly governed by crown cover and the influence of shade of the fast developing ground vegetation. In this project a direct influence of disturbance on light radiation on the ground was found only for the first 2 years while the effect on temperature in 5 cm soil depth could still be found in summer 1993, the third year after disturbance. In similar gap experiments with beech stands on acidic/podsolic brown earth in the Solling area/Germany Bauhaus (1994) found lower mean soil temperature differences in the gap compared to control. He pointed out that differences in maximum temperatures were more pronounced and that the effect of maximum temperatures might be more important than means for nitrification.

No strong acidification pulses in soil solution could be shown so far. High base saturation of the cation exchange sites (around 60 % in upper horizons, around 90 - 100 % in deeper soil layers > 40 cm) seems to be a guarantee for rapid buffer reactions against acidity produced by N-mineralisation.

Although no acidification pulse could be found, chronic acidification by loss of base cations up to 8.5 $kmol_c$ C_b ha^{-1} yr^{-1} in the disturbed plot could be shown. Leaching of base cations obviously is already ongoing in the reference area. This means that in the reference area the last step of NO_3^--leaching out of the soil has been already reached. Studies of the element budgets of an old beech stand, a few kilometers away from our stands, under static site conditions have been conducted by Jochheim (1992) and Eichhorn (1995). Site conditions were nearly the same. For 1990/1991 Eichhorn found leaching rates of 0.81 $kmol_c$ NO_3^--N ha^{-1} yr^{-1} under the root zone: Leaching rates of 0.91 $kmol_c$ NO_3^--N ha^{-1} yr^{-1} have been found in this project. This means that N-oversaturation could be confirmed by flux rates in two cases for basalt bedrock sites. In the gap area this process is even accelerated by excessive NO_3^--leaching, induced by gap felling, leading to NO_3^--contamination of groundwater. Bauhus (1994) found in his gap experiments on acidic soils in the Solling area the following NO_3^--N leaching rates under the root zone in 1991: control: 0, gap: 91.7, limed gap (with ground vegetation cover): 34.1 kg N ha^{-1}. These leaching rates have been measured for the first year after disturbance. Long term observations have not yet been published. The flux rates have been measured at the centre of the gap with sparce vegetational cover only, where neither root uptake by surrounding trees nor by ground vegetation plays a role. In a limed gap, where remarkable ground vegetation biomass was found, nearly 1/3 of the NO_3^--N leaching rates compared to the unlimed gap were found. In these findings the importance of vegetation as an transient store for elements, especially for N, becomes visible.

The importance of transient store in ground vegetation and understory biomass can be pointed out by the findings in Table 3.2-5. The data from a harvest in August 1994, 3rd vegetation period after disturbance of canopy structure, show the fast developing biomass rates for gaps with gain in light radiation (11,700 kg biomass ha^{-1} in the centre of the gap). Most of the biomass is formed by *Sambucus racemosa* within perennial compartments.

Assuming that at the beginning of the project biomass- and N-storage in all subplots and all compartments was the same, and also keeping in mind the uncertainties shown as high standard deviations, some rough conclusions can be drawn: When looking at N-storage in different compartments, it is quite obvious, that a decrease in N-content in the mineral soil by N-mineralisation and leaching is partly compensated by uptake in vegetation biomass and storage (via litterfall) in the organic layer. At the centre of the gap, 17% of N is stored in above ground biomass, while a lower N storage (around 20%) was shown for the centre of the gap compared to the control plot. Mrotzek (1998) and Eichhorn (1995) found a maximum store of 70 kg N ha^{-1} in their research area (district 14) with similar site conditions compared to the control area of tis project. In district 14, ground vegetation is dominated by *Urtica dioica* and *Mercurialis perennis* (biomass of the ground vegetation: max. 2,500 kg/ha). Analyses of seasonal variation showed a maximum N-storage in June and August.

Part of the N stored in ground vegetation biomass (esp. in leaves) will be returned to the organic layer, as indicated as high N- store in organic layer of the centre and edge of the gap. The organic layer will be subject to mineralisation within some months or a few years. The store of N in root/rhizome biomass is of minor importance compared to total N-store and flux rates within the ecosystem as less than 2% of total N is stored in these compartments.

Gaseous losses of nitrogen by exhalation have been investigated by Brumme et al. (pers. communication 1999) in the control and gap area in the period of June 1991 until May 1992. In the gap area he found N_2O-N-losses of 0.22 kg ha^{-1} and in the control area 0.36 kg ha^{-1} respectively. CO_2-C losses of 3.7 t ha^{-1} in the gap and 3.8 t ha^{-1} in the control have been found. Comparing N_2O losses with flux rates and soil balances, no major influence of gaseous fluxes on total N-budgets can be determined. Nevertheless, Brumme et al. (1999) found that under anoxic conditions, which may occur on tracks used for harvesting procedures, much higher emission rates of N_2O, being relevant as a greenhouse gas, may occur.

As the weathering rate of base cations of basalt bedrock will be in the range of 1-2 kmol$_c$ ha^{-1} yr^{-1} (Nilsson and Grennfelt 1988), it is quite obvious that the actual input of acidity apart from N compounds already consumes the buffer capacity deriving from weathering processes.

When looking at N inputs and transformations in the undisturbed plot, critical loads for nitrogen, as defined above, are already exceeded. Net balances for the disturbed plot show that under these conditions additional stress can arise simply by changes in abiotic environmental conditions other than N-input rates, affecting the N cycle. These data show that dynamic factors (such as breakdown of forest ecosystems, thinning and harvesting procedures, thinning of crown structure by insects or by breaking of branches by hoar) must be taken into account when discussing critical loads for N (see also Gundersen and Rasmussen, 1988).

In this project soil warming effects have been produced by sudden events. As temperature change due to global warming takes place in the long term over decades, different effects on forest ecosystems can be expected. e.g. buffering ability of forest ecosystems will be different within a longer period. On the other hand the degeneration of a N-saturated forest ecosystem due to air pollution can be accompanied or even accelerated by global warming effects. So influences by sudden events must be expected. Following this idea the data from this project can be taken for maximum risk assessments of forest decline effects and/or global warming effects in N-saturated systems.

From the silvicultural point of view it is quite obvious that in N-saturated forest ecosystems even small gap felling will be accompanied by processes of excessive mineralisation and NO_3^--leaching. The consequences of this process will lead to groundwater contamination and will favour nitratophile ground flora. Natural regeneration of deciduous trees such as beech will be inhibited. Therefore silvicultural practices should try to avoid excessive mineralisation by maintaining a continuous canopy closure. Single stem harvesting, forming multi-aged stands with high diversity in canopy structure will be the best silvicultural practice for management of N-saturated forest stands.

Acknowledgements

I thank the German Federal Ministry of Research and Technology (BMBF) for financial support of the project B2 Za. Dr. J.-U. Winter, A. Reinhard and P. Möller I owe a debt gratitude for technical and laboratory support.

References

Ågren GI, Bosatta E (1988) Nitrogen saturation of the terrestrial ecosystem. Environ Pollut 54:185-197

Bauhus J (1994) Stoffumsätze in Lochhieben. Ber Forschungszentrum Waldökosysteme, Göttingen, Reihe A, Bd 113

Breemen N v, Mulder J, Driscoll, CT (1983). Acidification and alkalinization of soil. Plant and Soil 75: 283-308

Brumme R, Borken W, Finke S (1999) Hierarchical control on nitrous oxide emission in forest ecosystems. Global Biogeochemical Cycles 13:1137-1148

Cole DW, Miegroet Hv, Foster NW (1992) Retention or loss of N in IFS sites and elevation of relative importance of processes. In: Johnson D W and Lindberg S E (eds) Atmospheric deposition and forest nutrient cycling. Springer, Berlin:196-199

Driscoll CT, Schaefer DA (1989) Overview of nitrogen processes. In: Malachunk JL, Nilsson J (eds) The role of nitrogen in the acidification of soils and surface waters. Nordic Council of Ministers NORD 1989 92

Eichhorn J (1995) Stickstoffsättigung und ihre Auswirkungen auf das Buchenwaldökosystem der Fallstudie Zierenberg. Ber Forschungszentrum Waldökosysteme, Göttingen, Reihe A, Bd 124

Eichhorn J, Paar U (1996) Deposition rates in open field. Personal communication, Hessische Forstliche Versuchs- und Forschungsanstalt, Hann. Münden

Grennfelt P, Thörnelöf E (1992) Critical Loads for Nitrogen – a workshop report. Nordic Council of Ministers working group 1 NORD 1992 14

Gundersen P, Rasmussen L (1988) Nitrification, acidification and aluminium release in forest soils. In: Nilsson J and Grennfelt P (eds) Critical loads for sulfur and nitrogen. Report from a workshop held at Skokloster Sweden 19-24 March 1988 NORD 1988 15

Jochheim H (1991) Chemische Bodeneigenschaften der Fest- und Lösungsphase in einem Buchenwaldökosystem in der Phase der Humusdisintegration. In: Eichhorn (ed) Fallstudie Zierenberg: Streß in einem Buchenwaldökosystem in der Phase einer Stickstoffübersättigung. Forschungsber Hessisches Ministerium für Landesentwicklung, Wohnen, Landwirtschaft, Forsten und Naturschutz 13, Wiesbaden:20-25.

Lischeid G (1995) Prozessorientierte hydrologische Untersuchungen am Kleinen Gudenberg bei Zierenberg (Nordhessen) in verschiedenen Skalenbereichen. Ber Forschungszentrum Waldökosysteme, Göttingen, Reihe A, Bd 128

Meiwes KJ, Hauhs M, Gerke H, Asche N, Matzner E, Lamersdorf N (1984) Die Erfassung des Stoffkreislaufs in Waldökosystemen. Ber Forschungszentrum Waldökosysteme, Göttingen, Bd 7:68-142

Miegroet H v, Cole DW, Foster NW (1992) Nitrogen distribution and cycling. In: Eichhorn J (ed). Atmospheric deposition and forest nutrition cycling – A synthesis of the integrated Forest Study (W. Johnson and S.E. Lindberg, eds.) Springer, Heidelberg:178-195

Mrotzek R (1998) Wuchsdynamik und Mineralstoffhaushalt der Krautschicht in einem Buchenwald auf Basalt. Ber Forschungszentrum Waldökosysteme, Göttingen, Reihe A, Bd 152

Nilsson J (1986) (ed) Critical loads for nitrogen and sulphur. Miljörapport 1986, II. Copenhagen

Nilsson J, Grennfelt P (1988) Critical loads for Sulfur and Nitrogen. Report from a workshop held at Skokloster Sweden 19-24 March 1988, NORD 1988 15

Skeffington RA, Wildson EJ (1988) Excess Nitrogen deposition: issues for consideration. Environ Pollut 54:159-184

Ulrich B (1983) Interaction of forest canopies with atmospheric constituents: SO_2, alkali and earth alkali cations and chloride. In: Ulrich B, Pankrath J (eds) Effects of accumulation of air pollutants in forest ecosystems. D Reidel Publishing Comp, Dordrecht:33-45

Vitousek PM, Gosz JR, Grier CC, Melillo JM, Reiners WA (1982) A comparative analysis of potential nitrification and nitrate mobility in forest ecosystems. Ecological Monographs 52:155-177

4 Forest restoration on degraded sites

4.1 Amelioration of an acid forest soil by surface and subsurface liming and fertiliser application

K. J. Meiwes

Forest Research Institute of Lower Saxony, Grätzelstraße 2,
D-37079 Göttingen, Germany, e-mail: meiwes@nfv.gwdg.de

Abstract

In Central Europe many forest soils are highly acidified. In the future management of forests concepts have to be developed as to, how reclamation of these soils can be achieved. Liming and fertilizing is one option, which may be combined with the selection and planting of site-specific tree species and with the establishment of an appropriate soil vegetation and shrub layer.

A field experiment in the Solling region, Germany, is presented, where a highly acid soil (Dystric Cambisol) was limed and fertilized to study the effects on soil chemical changes. After a clear cut of the former spruce stand, lime was applied either on the soil surface at a dose of 4 t ha^{-1} (LIM) or was mixed (22 t ha^{-1}) with deeper soil by ploughing to ameliorate both the surface and the deeper soil acidity (LPF). In addition on LPF plot 50 kg K ha^{-1} and 105 kg P ha^{-1} were applied. European beech, Norway spruce and black alder were planted in mixed stands. Five years later soil solution concentration and soil solid phase were studied. In LIM pH, effective CEC, and exchangeable Ca and Mg increased in the litter layer. In the surface mineral soil (0-5 cm) of LIM only exchangeable Mg was higher. In LPF a base saturation of 40-50% was achived up to 70 cm depth. In the soil solutions from 20-30 cm depth alkalinity, Ca and Mg concentrations increased but Al concentration decreased in LPF, whereas in LIM Ca and Mg increase was the only change. At 1 m depth Ca in soil solution increased in LPF and Mg in LIM and LPF. Initially, mean NO_3 concentrations in the soil solutions of all treatments were 100-700 µmol L^{-1}. Afterwards, when dense vegetation cover had developed, NO_3-concentrations decreased to 10-100 µmol L^{-1}. It is concluded, that mixing sufficient quantitiy of lime with the soil by ploughing or other means can ameliorate acidity of the whole soil solum within a few years, whereas lime applied on the soil surface may need more time to become effective. Nitrate losses with seepage water may be minimized by establishing a dense vegetation cover.

Key words: cation exchange, base saturation, soil solution, calcium, magnesium, nitrate

4.1.1 Introduction

Forest soils of central Europe are highly degraded chemically, and as estimated by Vanmechelen et al. (1997), 25% of the forest soils can be regarded as highly acid. The high level of soil acidity is partly due to the history of the land use, associated with excessive exploitation for a long time in the past, and partly due to the accentuated acidification process that has occurred due to high atmospheric acid inputs during the last few decades. Acidification is associated with the loss of base cations from the soil that can lead to high growth risk by lowering the vitality of trees. In highly acid soils, the trees tend to produce flat root systems, with only restricted access to soil volume for the uptake of nutrients and water. The regeneration (amelioration) of such soils would therefore involve the neutralisation of soil acidity from the surface soil and the deeper soil depths. This will not be achieved by merely reducing acid inputs even over extended period of several decades, because buffering of acidity by natural processes of silicate weathering is a slow process. Moreover the amount of acidity buffered in most soils may equal the amount produced during nutrient use for future biomass production and removal from forest sites (Ferrier et al. 1995) and the buffering of the additional acid input by silicate weathering in soils may not occur.

Amelioration of the highly acid soils in the future management of forests will therefore include the application of materials like lime. Application of liming material to soils increases their acid neutralisation capacity. In Germany this practice of lime application to forest soils has been extensively followed since mid 80's. Commonly recommended amount of lime application to highly acid forest soils is about 3 t ha^{-1}, which is commonly applied on the soil surface in a forest stand. This lime application may be repeated every 10-15 years. Lime applied on the surface of a soil dissolves at a slow rate and its solution products, Ca, Mg and HCO_3 ions, move slowly into the soil. This means that the neutralisation of soil acidity occurs only at a slow rate when lime is applied on the soil surface. In order to enhance the solubilisation of applied lime it should be mixed into the soil.

Amelioration of an acid soil would include many other practices in addition to the application of lime to improve the chemical status of soils. It may include ecological based practices to increase the detritus decomposition such as the establishment of a suitable undergrowth of tree species and the development of a shrub layer. For example, growing lupins in forest stands improved soil properties (Hetsch and Ulrich 1979) which resulted from an increase in detritus decomposition and thus improving 'system internal' nutrient cycling processes.

The aims of this study were to compare the various methods of applying lime to a highly acid soil. Lime was applied either on the soil surface or was mixed with the soil to deeper depths. Soil chemical changes that occurred during the first five years will be presented. In other chapters the effects of these soil treatments on the growth of different tree species (Dohrenbusch et al. 2002) and of the soil vegetation (Roloff and Linke 2002) are presented.

4.1.2 Materials and methods

The study was carried out on a forest site in Solling, Germany, located at an altitude of about 500 m. Average annual precipitation at the site is about 1000 mm and the average annual temperature about 6.5 °C. The soil is classified as Typic Dystrochrept (USDA) or Dystric Cambisol (FAO). It has developed from a loess layer of 40-60 cm depth, which overlays weathered sandstone. The soil has low pH between 3.0 and 4.2 in $CaCl_2$, high exchangeable Al (> 85% of exchange sites), high organic matter content (about 5%) in the surface soil, and the silt fraction as dominating soil fraction. The clay minerals consist mainly of illites and vermiculites, with a small fraction of kaolinites and chlorites.

4.1.2.1   Site preparation

The site carried a 130 years old spruce stand, which was clearfelled in autumn 1989. In the summer 1990 the slash was removed from the site by racking and the following treatments each covering an area of about 1 ha were set up:
CON - Untreated control.
LIM - Application of 4 t ha^{-1} dolomitic lime on the soil surface (lime composition: Ca - 228 mg g^{-1} and Mg - 134 mg g^{-1}).
LPF - After the removal of tree stumps, 18.5 t ha^{-1} of carbonatic lime (lime composition: Ca - 440 mg g^{-1} and Mg - 8 mg g^{-1}) was applied, and the site was ploughed to a depth of 70 cm. An additional 4 t ha^{-1} of dolomite (composition as above), and 50 kg K ha^{-1} and 105 kg P ha^{-1} were applied on the soil surface and were worked in to the soil by using a disk harrow. K was applied as K_2SO_4, and P as partially acidulated rockphosphate.

In the spring of 1991, a chess board design was used to divide each treatment (CON, LIM, LPF) into 15 m x 15 m blocks on which either beech (*Fagus sylvatica* L.) or spruce (*Picea abies* Karst.) was planted. In addition on each of the spruce and beech blocks of all the three treatments (CON, LIM and LPF) black alder (*Alnus glutinosa* (L.) Gaertn.) was planted (for details see Dohrenbusch et al. 1999, 2002). Black alder is regarded as a cover tree to protect seedlings, especially of beech, from frost damage.

4.1.2.2   Soil sampling

Soil samples were collected from the 130-year old spruce stand prior to its clearfelling. The samples were collected from 64 points of a 25 m x 25 m grid systematically laid on the whole site. The following litter horizons and soil depths were collected: $L+O_F$, O_H, 0-5 cm, 5-10 cm, 10-20 cm, 20-30 cm, 30-50 cm, 50-70 cm, 70-100 cm. Soils from all the 64 points were used to measure soil pH, whereas for other soil analysis samples from only 15 sites were used. The 15 sites were systematically selected as every 4th sample site of the 64 sites described above.
In 1995/96 soil samples were collected by following a systematic sampling method. In each treatment every 4th or 5th 15 m x 15 m block was selected for sam-

pling to give 7 sampling points per treatment. In the center of each selected block and at 3 m apart on two opposite sides of the center 3 individual samples were collected. The samples of each sampling point were mixed to obtain one composite for each depth for analysis. Mineral soil samples of CON and LIM were collected by using a soil auger of 8 cm diameter to a depth of 30 cm. Samples from depths deeper than 30 cm (to 100 cm) were obtained by collecting two soil cores using Pürkauer soil corer. Litter layers from CON and LIM areas were collected from an area of 500 cm^2. For collecting soil samples from the LPF site an auger of 10 cm diameter was used for 0 cm to 100 cm soil depth.

For measuring chemical changes in the soil solution, 20 lysimeter cups were installed on each treatment (type: P80, Manufacturer: State Porcelain Manufacturer, Berlin). The selection of the locations to install lysimeters was based on a systematic approach whereby important site characteristics were covered and a sufficient number of replicates was included for sampling. Every installation place had one lysimeter cup installed in the surface soil, and a second cup in the subsoil at 100cm depth. Depth of cup installed in the surface soil varied, 20cm for CON and LIM treatments and 30cm for LPF treatment. Each lysimeter cup was connected to a vacuum flask which was placed in a bucket with a lid. The bucket was kept in a pit dug into the soil. Every time before the water sample was collected, the vacuum flask was evacuated to 0.4 bar using a hand pump and left for a week. The lysimeter samples were collected periodically from December 1991 to October 1996.

4.1.2.3   *Chemical analysis*

Soils were air dried and analysed following the methods described by König und Fortmann (1996a, b, c). Soil pH was measured in $CaCl_2$ suspension; the exchangeable cations were extracted by percolating with 1 N NH_4Cl the mineral soil and with 0.2 N KCl the litter layer samples. KCl extractable Ca and Mg were measured in the limed litter samples by using a sequential percolation method. This allowed to distinguish Ca and Mg from the exchange sites to those from dissolution of unreacted lime. C and N were measured using Heraeus Element Analyser. Cations in the NH_4Cl and KCl extracts and in the lysimeter solutions were analysed using ICP emission spectroscope. Chloride was measured using ferricyanide method, ammonium using Na- salicylate and Na-dichlorocyanurate, nitrate using the cadmium reduction method and phosphate using ammonium molybdate. N_{org} was oxidised using UV and was determined using hydrazine and sulfanilamide. DOC was measured with a Shimadzu 5050 analyser. In the soil solution sulphate (mg S L^{-1}) was calculated from the total S (S_t- mg L^{-1}) measured by using ICP and dissolved organic C (C_{org}- mg L^{-1}) in soils by using the following relationship: SO_4 -S $= S_t - C_{org}/130$. This relationship was obtained by analysing 2000 samples where sulfate was analysed by using ion chromatography (König, personal communication). Carbonate in the litterlayer was determined using a volumetric calcimeter.

4.1.3 Results and discussion

4.1.3.1   Test of site homogeneity

Soils collected under the previous spruce stand were used to test if there were any significant spatial differences in soil properties on areas designated for the three treatments. No statistical significant differences (Scheffé test, p <0,05) were observed among various areas for soil pH, total C_{org}, total N, total K, Ca and Mg in the litter layer, and exchangeable cations in the mineral soil (data not shown). However effective cation exchange capacity (CEC_e) in 0-5 cm soil and pH values in the litter layer were statistically different for the three treatment areas. In areas assigned for CON treatment soil had CEC_e of 154 mmol$_c$ kg^{-1}, for LIM 119 mmol$_c$ kg^{-1}, and for LPF 123 mmol$_c$ kg^{-1}. These differences were associated with C content in the mineral soils and were noted for future interpretation of the data.

In the O_H-horizon of the litter layer mean pH($CaCl_2$) values were 2.9-3.0 on areas assigned for CON and LIM treatments and were significantly different from the mean value of 3.2 observed in areas assigned for LPF treatment. This difference in pH values was not considered to be of any major consequence, because the application of 22.5 t ha^{-1} of lime to LPF treatment was expected to affect the base/acid status of this soil in a major way with expected pH changes of a much bigger magnitude.

4.1.3.2   Changes in litter layer

Forest management practices of clearcutting the stand and lime application are expected to affect the C and N contents of the litter layer. However the measurements taken 6 years after clearcutting and lime application did not show any significant difference in their amounts when samples from CON treatment were compared with those taken prior to clearfelling from areas designated for CON and LIM treatments (Table 4.1-1). In a similar comparison for N between soils from LIM treatment after 6 years showed no difference (T-Test, p < 0.05) to soils collected from CON and LIM designated area prior to clear cut. It may however be noted that the mean values for the amount of C and N showed a high spatial variability with the confidence limits for the mean lying between 10 to 17%, which would make it difficult to detect any changes of smaller magnitude.

Table 4.1-1. Organic carbon in the surface organic layer before clearcut (1989) and six years (1996) after applying treatments, CON – untreated (no lime applied), LIM – lime applied in 1989/90 on soil surface (without mixing), (mean values and 95% confidence intervals).

Treatment	n	Carbon t ha^{-1}	Nitrogen kg ha^{-1}
Before clearcut	10	46.1 ($\pm$ 6.1)	2183 ($\pm$ 383)
CON (1996)	7	54.1 ($\pm$ 8.8)	2420 ($\pm$ 361)
LIM (1996)	7	47.4 ($\pm$ 6.3)	2308 ($\pm$ 312)

Litter layer of LIM contained 187 $\pm$85 (std. dev.) kg ha^{-1} of undissolved dolomite [CaMg(CO$_3$)$_2$·], which was about 5% of the amount applied. On liming the pH(CaCl$_2$) of the litter samples increased from 3.2 to 4.0. There was also an increase in the effective cation exchange capacity (CEC$_e$) of the litter layer from 83 mmol$_c$ kg^{-1} to 198 mmol$_c$ kg^{-1}. This increase in CEC$_e$ on liming was linearly related to the increase in pH (Fig. 4.1-1). Similar relationship was observed by Meiwes et al. (1996) for a nearby beech stand (*Fagus sylvatica* L.), but with the regression line of a slightly different slope (CEC$_e$ = 112 x pH - 134, r^2 = 0.76 for the beech stand compared to CEC$_e$ = 105 x pH - 210, r^2 = 0.58 in this study). The increase in CEC$_e$ was accompanied by an increase of 80 mmol$_c$ kg^{-1} exchangeable Ca and of 66 mmol$_c$ kg^{-1} exchangeable Mg (Table 4.1-2), a ratio of 1.21 for Ca:Mg. However, in the dolomite applied in LIM treatment, the Ca:Mg ratio expressed on mol$_c$ basis was 1.03. A congruent dissolution of dolomite would have released similar amount of Ca and Mg into the solution. A slightly higher fraction of Ca on the exchange sites of the litter layer indicated that Ca was more strongly retained by the exchange complex than Mg. In the LIM treatment exchangeable Ca in the litter layer increased by 2.9-fold and Mg by 7.4-fold.

Table 4.1-2. Exchangeable calcium and magnesium in the litter layer in the control (CON) and lime (LIM) plots (n = 7), 6 years after treatments were applied.

Treatment	Ca		Mg	
		mmol$_c$ kg^{-1}		
	Mean	Std.-dev	Mean	Std.-dev
CON	41.7	17.9	10.3	4.4
LIM	122.1	40.8	76.5	29.9

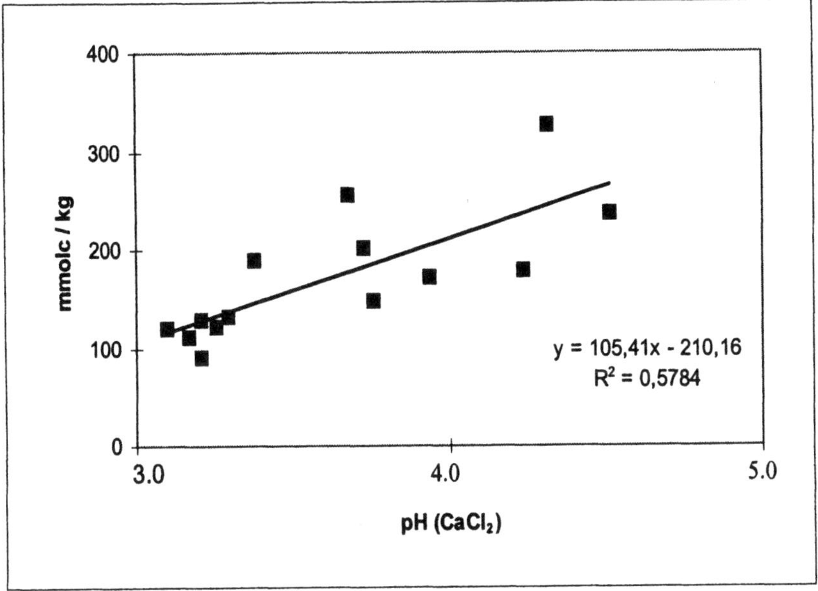

Fig. 4.1-1. Effective cation exchange capacity as a function of pH(CaCl$_2$) in the litter layer.

4.1.3.3   Changes in mineral soil

Solid phase

Application of lime on the soil surface (LIM treatment) changed the pH of the 0-5 cm soil only, and no effect was observed in the soil at deeper depths (Fig. 4.1-2). However mixing of lime with the soil at the surface and at deeper depths (LPF treatment) showed a uniform increase in soil pH to values of more than 4.3. A soil pH value 4.2 in the salt solution is considered to be critical below which the Al starts appearing in substantial and measurable amounts in the soil solution. LPF treatment can therefore be considered to have changed the soil pH to desired level. Other major changes in soil chemical properties were also observed in the LPF treatment, for example, organic C and total N were lower in the surface soil but higher in the subsoil (30-70 cm depth) from LPF treatment than in those from CON and LIM treatments (Fig 4.1-3).

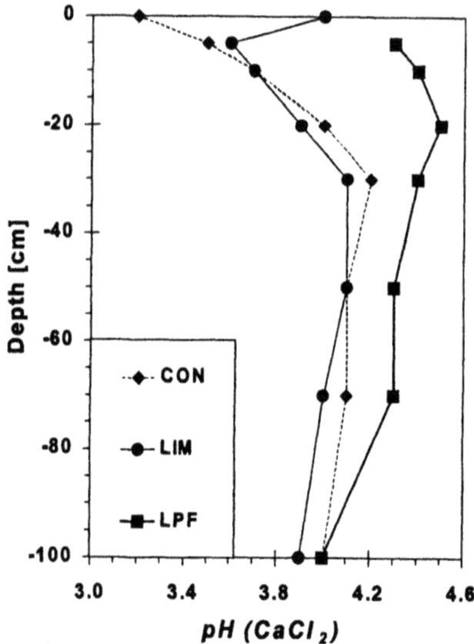

Fig. 4.1-2. pH values in different soil depths 6 years after following treatments were applied: untreated – CON, limed on the soil surface – LIM, and limed mixed with ploughing and fertilised – LPF.

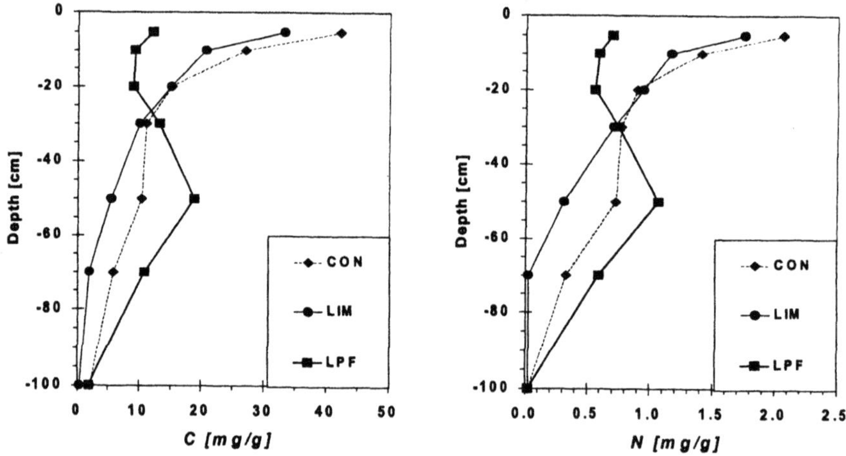

Fig. 4.1-3. C and N in the mineral soil at different depths 6 years after following treatments were imposed: untreated – CON, limed on the soil surface – LIM, and limed mixed with ploughing and fertilised – LPF.

CEC$_e$ values ranged from 90 to 120 mmol$_c$ kg^{-1} in the surface 10 cm soils of CON and LIM treatments (Table 4.1-3), and were lower in the deeper soil layers (50-70 mmol$_c$ kg^{-1}). This depth gradient in CEC$_e$ values was not observed in the LPF treatment because of the soil mixing by ploughing. However the equivalent fraction of exchangeable calcium on the exchange complex (X^S_{Ca}) increased in the LPF treatment from about 1-5 % of CEC$_e$ (CON) to 30-40% in 0-70 cm depth, and from 1% to 9% of CEC$_e$ in 70-100 cm soil depth. This increase of X^S_{Ca} in the soil below the ploughed layer can therefore be assigned to the transport of Ca via soil solution.

Equivalent fraction of exchangeable Mg (X^S_{Mg}) increased from about 1% of CEC$_e$ in CON treatment soil at 0-30 cm depth to 14-18% in LPF. On account of K application in LPF treatment the equivalent fraction of potassium (X^S_K) increased from 1-2% of CEC$_e$ (CON) to 3-7% of CEC$_e$ in the surface 20 cm soil (data not shown).

In the LIM treatment surface application of 4 t ha^{-1} of dolomite resulted in only a slight increase of X^S_{Ca} from 5 to 8% of CEC$_e$ at 0-5 cm depth, which however was statistically not significant. The increase in Mg was more pronounced; X^S_{Mg} increased from 1% to 4-6% at 0-10 cm depth. This increase in Mg was evident at even 10-30 cm soil depth. Significantly higher values of X^S_{Mg} were noted in soils from LPF treatment than those from CON and LIM treatment. The difference in CEC$_e$ between CON and LIM was not just a treatment effect, but also a site effect, which occurred before the experiment started, therefore those results have to be interpreted with caution.

Table 4.1-3. Mean values of CEC$_e$ and equivalent fractions (X^S) of Ca and Mg five years after application of dolomite on the soil surface (LIM) and after liming, ploughing and fertilisation (LPF), (CON = Control), the different letters point to significant differences in mean values (p < 0.05, Scheffé test; forest floor: Mann Whitney U test).

Depth	Treatment			Treatment			Treatment		
	CON	LIM	LPF	CON	LIM	LPF	CON	LIM	LPF
cm	CEC$_e$ (mmol$_c$ kg^{-1})			X^S_{Ca}	(% of CEC$_e$)		X^S_{Mg}	(% of CEC$_e$)	
Forest floor	83 [a]	198 [b]		44 [a]	55 [b]		11 [a]	34 [b]	
0 - 5	123 [a]	103 [b]	59 [c]	5 [a]	8 [a]	32 [b]	1 [a]	6 [b]	14 [c]
5 - 10	101 [a]	86 [a]	54 [b]	3 [a]	5 [a]	32 [b]	1 [a]	4 [a]	17 [b]
10 - 20	67 [a]	63 [a]	54 [b]	2 [a]	3 [a]	35 [b]	1 [a]	2 [a]	18 [b]
20 - 30	42 [a]	48 [b]	61 [b]	1 [a]	3 [a]	38 [b]	1 [a]	2 [a]	16 [b]
30 - 50	51 [a]	45 [a]	79 [b]	2 [a]	2 [a]	46 [b]	1 [a]	1 [a]	6 [b]
50 - 70	51 [ab]	49 [a]	61 [b]	2 [a]	1 [a]	37 [b]	1 [a]	1 [a]	4 [b]
70 -100	53 [a]	47 [a]	51 [a]	1 [a]	2 [a]	9 [b]	1 [a]	1 [a]	1 [b]

Soil solution

In the lysimeter solutions collected from the surface soil in May 1996, Ca was the dominant cation in the LPF treatment (Table 4.1-4), with its equivalent fraction lying around 54% of the sum of cations in the solution. Mean concentration of Al in the solution was 8 µmol L^{-1}, which accounted for 4% of the cation sum. These Ca and Al values in the soil solution corresponded to a soil, which would usually have about 50% of its exchange sites occupied by base cations. The mean concentration of Mg in the soil solution of LPF treatment was 52 µmol L^{-1}, which was three times higher than that found on the control plot.

Table 4.1-4. Mean concentration of elements in the soil solution collected with lysimeters from the surface soil (May 1996) and subsoil (Oct. 1996), the different letters point to significant differences in mean values (Kruskal Wallis test; the differences were located by using Mann Whitney U test p < 0.05)

Element	Surface soil			Subsoil		
	CON	LIM	LPF	CON	LIM	LPF
		µmol L^{-1}			µmol L^{-1}	
Na	95.6 [a]	89.3 [a]	142 [a]	87.7 [a]	85.7 [a]	71.5 [b]
K	12.6 [a]	13.9 [a]	24.7 [a]	28.1 [a]	19.3 [a]	26.2 [a]
NH_4	7.2 [a]	5.0 [a]	5.0 [a]	5.0 [a]	5.0 [a]	6.2 [a]
Ca	31.8 [a]	52.8 [b]	173 [c]	50.0 [a]	50.9 [a]	218 [b]
Mg	14.4 [a]	72.0 [b]	52 [b]	17.9 [a]	35.3 [b]	28.7 [b]
Mn	25.1 [a]	10.4 [b]	3.5 [c]	23.4 [a]	17.6 [b]	9.5 [c]
Al	28.4 [a]	37.9 [a]	7.7 [b]	96.8 [a]	71.2 [ab]	52.7 [c]
Cl	79.4 [a]	87.1 [a]	91.3 [a]	42.4 [a]	70.5 [b]	45.5 [a]
NO_3	55.3 [a]	176 [b]	52.8 [a]	44.1 [a]	71.6 [a]	14.5 [b]
PO_4	14.1 [a]	5.0 [a]	5.0 [a]	5.0 [a]	5.0 [a]	5.0 [a]
SO_4	104 [a]	114 [a]	176 [b]	298 [a]	229 [b]	381 [a]
N_{org}	5.5 [a]	11.9 [a]	10.6 [a]	1.2 [a]	4.2 [a]	0.1 [b]
DOC	603 [a]	730 [a]	1190 [a]	n.d.	n.d.	402.0 [a]
pH	4.6 [a]	4.6 [a]	5.0 [b]	4.4 [a]	4.5 [b]	4.6 [c]
Alkalinity ($µmol_c L^{-1}$)	-117 [a]	-132 [a]	127 [b]	- 430 [a]	- 326 [b]	- 230 [c]

n.d., not determined

In the LIM treatment Ca concentration in the lysimeter solutions collected from the surface soil had increased by 1.6 times (from 32 to 53 µmol L^{-1}) when compared to solutions from CON treatment (Table 4.1-4). However the mean Al concentrations remained similar in both treatments. The mean Mg concentration was slightly higher in LIM solutions than in those of LPF, despite both receiving a similar amount of dolomite in the surface soil. However the difference was not significant. During the winter 1991/92 mean NO_3 concentrations ranged from 185-980 μmol L^{-1} with highest and significantly different concentrations in LPF (data not shown). From autumn 1992 to 1995 mean NO_3 concentrations were in the range of 5-55 μmol L^{-1} with no treatment effects. In May 1996 NO_3 concentrations in LIM were higher than in CON and LPF (see Table 4.1-4). The high NO_3-concentrations at the beginning of experiment may be due to soil disturbances following clear cutting and also due to mixing of lime with the mineral soil in LPF. After one year NO_3 concentrations decreased and remained low without any difference between the treatments except in May 1996.

Solutions collected from the subsoil of LPF showed significantly high pH and high alkalinity (defined as: $Na + K + NH_4 + Ca + Mg - SO_4 - Cl - NO_3$) when compared with solutions from CON and LIM treatments. The three treatments could be ranked in the following order for alkalinity: CON < LIM < LPF. Concentration of Ca was significantly higher in the solution from the subsoil of LPF treatment than those from CON and LIM treatments. Concentration of Mg from both the LIM and LPF treatments was higher than from CON treatment, indicating dolomite, whether applied on surface of the soil or mixed with the surface soil, had similar effect on the increase in Mg in soil solutions to deeper soil depths. Contrary to Mg there was no effect on the concentration of K in the solutions from subsoils of LIM or LPF treatments (Table 4.1-4).

Acid output from these soils occurred primarily through the output of Al in the solution phase (Table 4.1-4). Concentration of Al (and also of Mn) in soil solutions decreased on liming (LPF and LIM treatments). In LPF treatment it was also the result of mixing high amounts of lime (22.5 t ha^{-1}) with the soil.

Among the anions sulphate was the important one representing 76-92% of the anion sum in equivalent values in the solution of the subsoil. Sulphate concentration was highest in LPF and lowest in LIM.

The changes in Mg concentration in the surface soil (Fig 4.1-4) point to the dynamics in the soil processes affecting the acid-base status of the soil. In LPF soil about one year after ploughing and liming (end of 1991) the Mg concentrations were very high, about 250 μmol L^{-1}. They decreased then continuously until the values were 40-60 μmol L^{-1} in the vegetation growing period of 1994.

In surface soil solutions from LIM treatment Mg concentrations were initially similar (about 35 μmol L^{-1}) to those of CON (Fig 4.1-4). Whereas in CON Mg concentrations decreased continuously to about 15 μmol L^{-1} by 1994 in the LIM treatment they remained constant around 30 μmol L^{-1} until 1994 but then started to increase. In 1996 Mg concentration in the solutions from LIM treatment was about 70 μmol L^{-1}. In LPF the mean concentrations of Ca decreased continuously from 600 μmol L^{-1} in 1991 to about 150 μmol L^{-1} in 1996. In solutions from CON and LIM the mean concentrations of Ca decreased from 70 μmol L^{-1} to 50 mmol L^{-1} (CON) and from 70 to 40 μmol L^{-1} (LIM) respectively (data not shown).

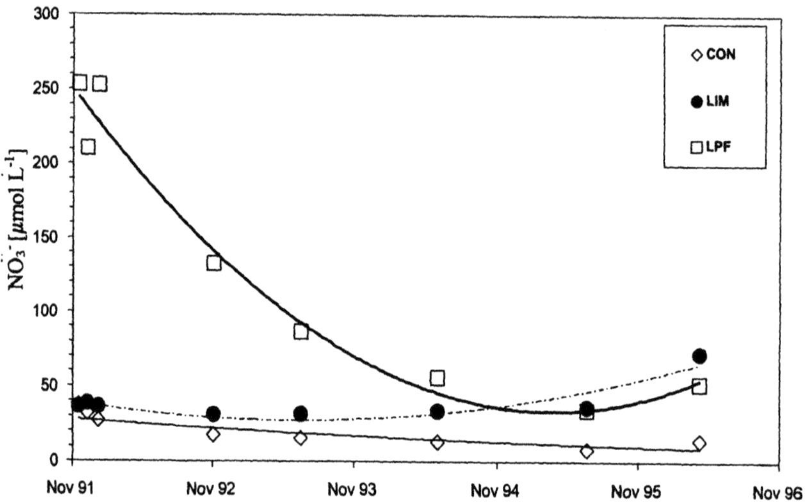

Fig. 4.1-4. Mean Mg concentrations in lysimeter solutions from the surface soil (treatments: untreated – CON, limed on the soil surface – LIM, limed, ploughed and fertilised – LPF).

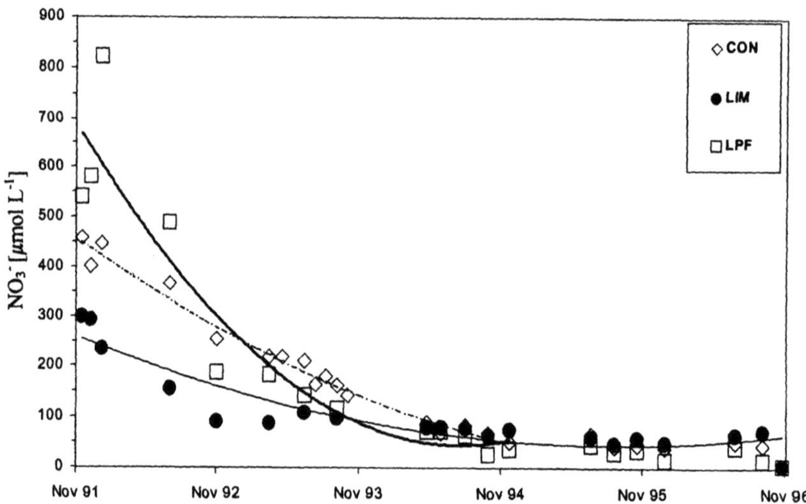

Fig 4.1-5. Mean NO_3 concentrations in lysimeter solutions from the surface soil (treatments: untreated – CON, limed on the soil surface – LIM, limed, ploughed and fertilised – LPF).

Mean concentrations of nitrate in solutions collected from 100 cm depth during the first three sampling periods (Dec. 91 to Feb. 92) (Fig. 4.1-5) were highly variable in all the three treatments, LPF (550-850 μmol L^{-1}), CON (400-500 μmol L^{-1}) and LIM (250-300 μmol L^{-1}). The mean values of nitrate were significantly different (Mann Whitney U test, P < 0.05) between CON and LIM for Dec. and Feb. samplings, and between LPF and LIM for Feb. sampling only. Thus the application of heavy doses of lime and soil disturbance (ploughing) did not change the nitrate concentration in LPF site in any significant way. Nitrate concentrations were low during the vegetation period of 1993 (100-300 μmol L^{-1}), whereas the differences between the treatments were still evident, though not significant. Since 1994 the average nitrate concentrations at 100 cm depth were similar on all treatments and ranged from 50-100 μmol L^{-1}.

4.1.4. Practical implications

Clearcutting of the spruce stand caused an increase in the nitrate concentration in the drainage solution at 100 cm. However the concentrations were lower than those reported by others in literature. For example, Kölling (1993) reported that 3 years after windthrow of spruce stands in Southern Germany, the mean nitrate concentrations were 600 μmol L^{-1} (std dev.: 300 μmol L^{-1}, n = 13 sites). Mixing of lime with the soil on the clearcut site led to an increase in nitrate concentrations and losses whereas application of lime on the soil surface caused lowering of nitrate concentration in the percolating water. This may be related to higher N uptake as was evident by higher biomass of soil vegetation in the LIM treatment than on the other two (CON and LPF) treatments (Roloff and Linke 2002, this volume). Similar type of relationship between nitrate concentrations and soil vegetation was reported by Bauhus and Bartsch (1995). They observed lower concentration of nitrate in vegetation gaps on sites receiving lime than on those where no lime was applied.

Changes of soil N content resulting from treatments could not be described by statistical means because of the variability observed both in the litter layer and mineral soil. Nykvist (1977) pointed out that the balance sheet approach to show changes in nutrient levels in soils can only be used if huge statistical differences are available. Therefore the use of nitrate concentration in soil solutions, when combined with the estimation of the quantity of water drained, can be a useful technique of assessing nitrate losses (Kölling 1993).

Application of lime on the soil surface can ameliorate soil acidity, but the amelioration occurs at a slow rate starting at the soil surface, which then proceeds to lower depth. This has already been reported in a number of other experiments (Meiwes 1994). By mixing the lime with the soil amelioration process occurs at a fast rate. Changes in the Mg and Ca concentrations in the surface soil give only an indication of the rate at which the lime dissolves in LPF treatment. It is possible that high concentration of Ca and Mg in the soil solution is associated with the phase of lime dissolution, which has completed during the 6-year period. This would translate into a mean dissolution rate of about 4 t ha^{-1} a^{-1}. Dissolution rates of lime applied on the soil surface are low usually less than 1 t ha^{-1} a^{-1} (Kreutzer 1995). Another important distinction between the lime applied on the soil surface and lime mixed with the soil is the time required to ameliorate most of the subsoil acidity. In the present study application of lime on the soil surface (LIM treatment) increased exchangeable Mg in mineral soil, but most of the Ca was retained on the exchange sites in the litter layer. Mixing of lime by ploughing the soil to 70 cm depth caused an increase in the Ca levels to 100 cm depth, i.e., 30 cm below the treated zone.

The effect of mixing lime by ploughing can also be demonstrated by comparing LPF with the results of a liming experiment on a similar site close to this study. In that experiment dolomite was applied on the soil surface under a beech forest at the rate of 30 t ha^{-1} (Beese 1989). After 11 years 0-5 cm soil had a base saturation of 50% of the CEC$_e$ and 5-10 cm soil only 20% (Meiwes 1995), while in LPF base saturation was 40% at 70 cm depth five years after mixing lime (22.5 t ha^{-1}) with the mineral soil.

4.1.5. Conclusions

1. Clearcutting of a spruce forest caused significant losses of soil N as nitrate leaching, but any management practice that can lead to fast establishment of the vegetation cover, will diminish these leaching losses. In the management of forest stands in Germany, the usual practice is to avoid liming on clearcut areas, because of the expected environmental hazard of producing high nitrate losses. However the results shown here indicate that this recommendation requires new examination for future considerations.
2. Mixing of sufficient quantity of lime with the soil by ploughing or other means can quickly within a few years ameliorate the acidity of the soil solum, whereas lime applied on the soil surface may need more time to become effective at deeper soil depths. Mixing lime with the soil provides the forest manager with a suitable silvicultural practice to ameliorate highly acid soils and to restore the functioning of forest ecosystems in an optimum manner. Such a silvicultural practice will include selection and planting of site specific tree species and the establishment of soil vegetation and shrub layer with an aim to increase the elasticity of soils to acidification, and to provide optimum supply of nutrients for tree growth.

Acknowledgements

I thank Mr. H. Bartens for the data handling and Dr. P.K. Khanna for the critical review of the manuscript. Funding was provided by the German Bundesministerium für Bildung, Wissenschaft, Forschung und Technologie (BMBF), 325-7291 OEF 2019.

References

Bauhus J, Bartsch N (1995) Mechanisms for carbon and nutrient release and retention in beech forest gaps. Plant Soil 168-169:579-584.

Beese F (1989) Wirkungen von Kalkungs- und Düngungsmaßnahmen auf die chemische Zusammensetzung der Bodenlösung. Ber Forschungszentrum Waldökosysteme, Göttingen, Reihe A, Bd 49:27-48

Dohrenbusch A, Kumke J, Mackenthun G (2002) Effects of soil amelioration on growth and nutrient supply of different tree species (this volume)

Ferrier RC, Wright RF, Cosby BJ, Jenkins A (1995) Application of the MAGIC model to the Norway spruce stand at Solling, Germany. Ecol Modelling 83:77-84

Hetsch W, Ulrich B (1979) Die langfristige Auswirkung von Kalkung, Bodenbearbeitung und Lupinenanbau auf die Bioelementvorräte zweier Flottsandstandorte im FA Syke, Forstw Cbl 98:237-244.

König N, Fortmann H (1996a) Probenvorbereitungs-, Untersuchungs- und Elementbestimmungs-Methoden des Umweltlabors der Niedersächsischen Forstlichen Versuchsanstalt und des Zentrallabors II des Forschungszentrums Waldökosysteme. Teil 1: Elementbestimmungsmethoden A-M. Ber Forschungszentrum Waldökosysteme, Göttingen, Reihe B, Bd 46

König N, Fortmann H (1996b) Probenvorbereitungs-, Untersuchungs- und Elementbestimmungs-Methoden des Umweltlabors der Niedersächsischen Forstlichen Versuchsanstalt und des Zentrallabors II des Forschungszentrums Waldökosysteme. Teil 2: Elementbestimmungsmethoden N-Z und Sammelanhänge. Ber Forschungszentrum Waldökosysteme, Göttingen, Reihe B, Bd 47

König N, Fortmann H (1996c) Probenvorbereitungs-, Untersuchungs- und Elementbestimmungs-Methoden des Umweltlabors der Niedersächsischen Forstlichen Versuchsanstalt und des Zentrallabors II des Forschungszentrums Waldökosysteme. Teil 3: Gerätekurzanleitungen und Gerätekurzanleitungen Datenverarbeitung. Ber Forschungszentrums Waldökosysteme, Göttingen, Reihe B, Bd 48

Kölling C (1993) Die Zusammensetzung der Bodenlösung in sturmgeworfenen Fichtenforst (*Picea abies* (L) Karst.) - Ökosystemen. Forstl Forschungsber München 133:1-134

Kreutzer K (1995) Effects of liming on soil processes. Plant Soil 168-169:447-470

Meiwes KJ (1994) Kalkungen. In: Matschullat J, Heinrichs H, Schneider J, Ulrich B (eds.) Gefahr für Ökosysteme und Wasserqualität - Ergebnisse interdisziplinärer Forschung im Harz. Springer-Verlag Berlin:416-431

Meiwes KJ (1995) Application of lime and wood ash to decrease acidification of forest soils. Water Air Soil Pollution 85:143-152

Meiwes KJ, Mindrup M, Khanna PK (1996) Changes in chemical parameters of humus layers under a beech forest following liming. In: Hovmand MF (ed.). Mineral Cycling and Air Pollution Fluxes in Reforested Heathland. Proceedings of the Intern. Workshop 26.-28. Aug. 1996. Ulfborg. National Environm. Res. Inst. DK 4000 Roskilde. Denmark:56-57

Nykvist N (1977) Changes in the amounts of inorganic nutrients in the soil after clearfelling. Silva Fennica 11:224-229

Roloff A, Linke J (2002) Agricultural herbs as auxiliary plants in stand establishment (this volume)

Vanmechelen L, Groenemans R, Van Ranst E (1997) Forest soil condition in Europe . EC-UN/ECE Brussels. Geneva:111-117

4.2 Effects of soil amelioration on growth and nutrient supply of different tree species

A. Dohrenbusch, J. Kumke, G. Mackenthun

Institute of Silviculture, Göttingen University, Büsgenweg 1,
D-37077 Göttingen, Germany, e-mail: adohren@gwdg.de

Abstract

A mixed stand of spruce and beech, including additional 8 tree species, was established in the higher elevations of the Solling (FA Neuhaus) (500 m above sea level, 1100 mm of precipitation). Different cultivation treatments were applied to the originally strongly acidified soils (pH 4, base saturation on the exchanger < 5% of the cation exchange capacity): a. (LPF) deep ploughing with very intensive liming (22.5 t ha^{-1}), fertilisation with 120 kg P_2O_5 ha^{-1}, b. (LIM) surface liming with 4 t ha^{-1} of bitter spar and, c. (CON) no soil treatment. Four to six years after stand establishment the following results can be summarised:

Mountain ash (*Sorbus aucuparia*), black alder (*Alnus glutinosa*), birch (*Betula pendula*) and Norway spruce (*Picea abies*) were the only species, that did not have greater losses. Extremely high rates of losses ranging from 70-90% were registered for noble fir (*Abies procera*), ash (*Fraxinus excelsior*) and Norway maple (*Acer platanoides*). The rates for sycamore maple (*Acer pseudoplatanus*) and douglas fir (*Pseudotsuga menziesii*) varied from 40-50%, which is in the average field but still not acceptable. This gradation of the plant losses can be reflected onto the growth of the remaining trees as well. The pioneer species had a mean height of 2.5 m six years after planting, the height for the demanding species was below the average height of one meter.

The impact of soil cultivation increases with the grade of demand of the tree species.

Four years after planting a significant increase in height growth, after the, in practice mainly used, surface liming with 4 t ha^{-1} of magnesium containing lime, was shown for sycamore maple in comparison to the control plot. Most trees showed an improvement of up to 20%, which could be not statistically proven though. The intensive amelioration (LPF) showed a significant improvement in height growth for all species. Especially the more demanding trees, such as maple and ash, showed a shot length of two to three times as long as for the control. These striking site-effects were visible for the valuable deciduous species two years after planting and increased significantly within the following years.

The diameter growth responded less to the improved site conditions. The relative increase in growth was less distinct than for height growth.

Soil amelioration creates favourable conditions for seeding, especially for willow (*Salix caprea*). The remaining species originated from natural seeding, such as birch, spruce and mountain ash grew in higher numbers with less favourable chemical site conditions.

Key words: afforestation, site preparation, amelioration, nutrient supply, initial growth, *Picea abies, Fagus sylvatica, Alnus glutinosa, Acer pseudoplatanus, Acer platanoides, Fraxinus excelsior, Abies procera, Betula pendula, Sorbus aucuparia*

4.2.1 Introduction

The concepts of evaluating and managing forests in Central Europe have undergone fundamental changes in recent years. The central silvicultural aim, is to obtain an approach of our, in terms of species and structure, man-influenced forests towards a potentially natural species composition. This is mainly achieved by distinctly supporting deciduous species.

Two different approaches are taken to realise this type of sylvicultural concept in practice. Conifer stands, that did not reach maturity yet, can be managed using a gradual change in tree generations or at least partially mixing in hardwood species using underplanting. The faster way to obtain this central aim though, is to establish the wanted forest after clear-cut. Even though the use of clear-cut as a management tool is mainly banned, it will still be present up to a certain extent. On top of that possibly increasing storm events due to climatic changes in Central Europe, lead to the need to recultivate the affected areas. The base of all these concepts is the exact knowledge of tree-specific site-conditions, regardless of the method applied for the alteration of species. A long-term afforestation research plot was constructed on a fully cleared, storm affected area in the higher elevations of the Solling. Besides a wide range of tree species, different types of soil cultivation were used to promote the cultures.

4.2.2 Methods and material

4.2.2.1 Research plots

The afforestation plots are located on the Solling-plateau, 500 m above sea level, within the logging units 4253 and 4255, Forest District Dassel. This area is part of the growth zone Oberer Solling, characterised by cool and humid climatic conditions. The mean annual temperature is 6.4 °C and 12.3 °C during the growing season from May to September. The long term mean value for annual precipitation is 1080 mm, half of which are measured during the growing season (measuring station Torfhaus, Solling).

According to measurements of air pollutants, contamination with SO_2 is very high in winter. Average concentrations of more than 100 µg m^{-3} were measured, this approximates to concentrations found in industrial centres. The ozone concentration during summer was found to be very high level of 100 µg m^{-3}. The average nitrogen content of the air amounted to more than 50 µg m^{-3}. The overall input of sulphur decreased within the last 35 years though. After a maximum of 100 kg ha^{-1} yr^{-1} in the mid seventies the sulphur input decreased to about 50 kg ha^{-1} year^{-1} at the beginning of the nineties and comes to around 30 kg today. The overall amount of nitrogen-deposition, consisting of equal parts ammonium and nitrate, increased from 30 to 40 kg ha^{-1} yr^{-1}during the same period of time.

A slightly podsolic, similigleyic cambisol (FAO) with a 50 to 60 cm high deposition of loess developed on top of the geological formation Triassic red sandstone. These exposed areas are, due to the deposition-rates, highly acidic. The very low pH($CaCl_2$) of 3 in the top soil layer increases to a pH value of 4 in the deeper layers. The mineral soil has therefor a small supply of only 5% of sodium, potassium and magnesium to the overall exchange capacity (Meiwes 2002, this volume).

After a wind throw in 1987 a big area was cleared, the remaining root systems and the slash was pushed on rampages. Clearing forks were used to keep the major part of the humus on the plots. In the same year three different types of soil preparations were used on three plots, each one hectare in size:

- LPF: Deep ploughing and mixing in 18.5 t ha^{-1} of carbonatic dolomite lime up to a depth of 70 cm, 4 t ha^{-1} were mixed into the upper soil up to a depth of 20 cm. A fertilisation with potassium (in the form of potassium-sulphate), 50 kg ha^{-1}, and phosphorus 105 kg ha^{-1}, was applied in addition to the liming procedure. This procedure is an extremely intensive form of amelioration, not realistically usable in practice. It was done for this experiment to achieve an ideal chemical soil condition (for details see Meiwes 2002, this volume p. 169).
- LIM: surface liming with bitter spar (4 t ha^{-1}, practice oriented amount).
- CON: control-plot, without any soil cultivation.

The whole research area was divided into square sub-plots, 225 m^2 each, in shape similar to a chessboard. In April 1991 spruce (*Picea abies* Karst., clone-mix, cuttings, 50-70 cm) in a formation of 3 m x 1 m alternating with beech (*Fagus sylvatica* L., origin Solling 2+0, 50-70 cm) in a formation of 1,5 m x 1 m (angle planting) were planted on these sub-plots.

Within the 3 different cultivation types ten sub-plots were chosen by chance (5 for beech, 5 for spruce) and additionally planted with the following tree species:

Number of subplots	Tree species	Origin/assortment
11	*Acer pseudoplatanus*	(Westd. Bergl. 1+1)
11	*Fraxinus excelsior*	(Westd. Bergl. Solling 1+2)
11	*Abies procera*	(Mary's Peak. 2+1)
11	*Pseudotsuga menziesii*	(Uslar 2+0)
2	*Sorbus aucuparia*	(Willingshausen 1+1)
2	*Acer platanoides*	(1+1)
2	*Alnus glutinosa*	(1+1)
2	*Betula pendula*	(1+1)

To protect the culture, a nurse crop was established, planting black alder (*Alnus glutinosa* 1.2-1.5 m high), in a formation of 4 x 2.5 m, on the whole area.

4.2.2.2   Methods

Real repetitions of the 3 differently treated areas could not be done, because the soil cultivation for this practice oriented afforestation was done extensively. It was therefore necessary to examine the initial situation. This was done by examining the soil solid phase; no difference was found between the variations (for details see Meiwes 2002, this volume). Ground vegetation developed slightly stronger on the LIM-plot with a cover degree of 45% than on the CON-plot (35%). A difference in species contribution was not found.

On 10 subplots of the variations CON, LIM and LPF all trees were marked. Tree height, diameter of the shoot base and damages were recorded annually from 1991 to 1994. In addition to the secondary and mix tree species 20 beech or spruce trees growing right next to the plots were permanently marked. This way 100 individuals were available for each cultivation variety (CON, LIM, LPF). After the third growing season all naturally regenerated tree species were additionally registered. Last measurements, for the moment, of the tree heights were taken following the growing season of 1996. The culture was at that time six years old with plants eight to nine years of age. Chemical composition of the needles and leaves were analysed in late summer of 1994 and 1996.

4.2.3 Results

4.2.3.1   Plant damages

Starting in the early summer of 1991 pest attacks became a problem. In May the noble firs were already severely damaged by the big brown weevil (*Hylobius abietis*), insecticides had to be used to save the culture. In the fall of 1991 damages were found on both soft wood species. Small damages on the young spruce shoots were done by *Phyllobius arborator*. *Phyllaphis fagi* damaged the leaves of the beeches and *Otiorhynchus niger* damaged the roots of the spruces. To prevent damages by *Clethrionomys glareolus, Microtus agrestis* and *Arvicola terrestris* bait-traps were spread over the whole area and controlled regularly.

4.2.3.2   Success of regeneration

Figure 4.2-1 significantly shows the difficult situation for most tree species on the untreated control-plots. Pioneer species like mountain ash, alder and birch as well as spruce can survive this normal condition without significant losses (< 15%).

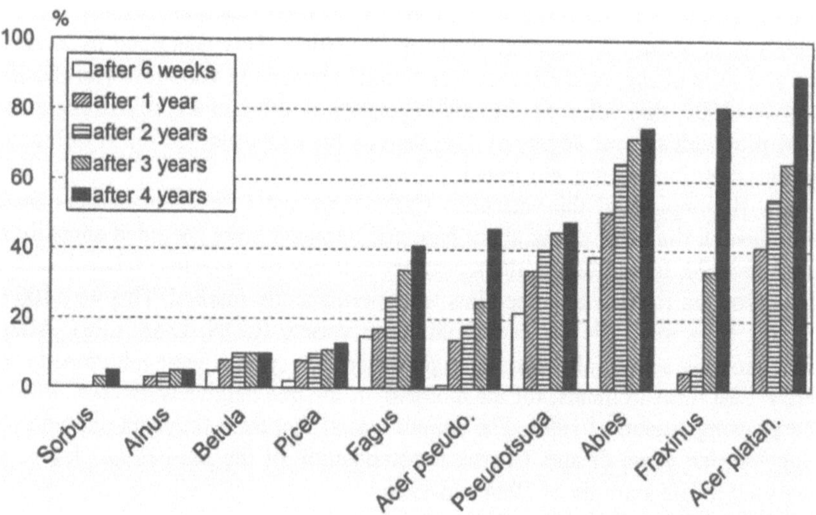

Fig. 4.2-1. Portion of plant losses during the first four years after planting on the control plot (CON).

Four years after stand establishment all other species had losses of at least 40%. Species specific differences were found within the first phase of stand development. Six weeks after planting a, compared to the further development, high number of beech, douglas fir and noble fir died. Following these losses right after planting, only a slight increase was detected for spruce within the following years. Even after the 6th growing season in October 1996, the average rate of 17% did not significantly change. Severe losses with every following year were observed for beech, mountain maple, douglas fir and noble fir. Even though the initial situation in terms of losses was very different for each species group (extremely high for fir), the losses proceeded at the same speed for all species in the following years. The development for Norway maple and ash was very different though. Ash had only 7% losses after two years, this dramatically increased to more than 80% after 4 years (Table 4.2-1).

Table 4.2-1. Percentage of plant losses (%) four years after stand establishment.
$\uparrow$ differences in contrast with the control plot (CON)
$\uparrow\uparrow$ distinct differences in contrast with the control plot (CON)

Species	Site preparation		
	CON	LIM	LPF
Sorbus aucuparia	5	0	0
Betula pendula	10	18	0 $\uparrow$
Alnus glutinosa	5	20	0 $\uparrow$
Picea abies	13	4 $\uparrow$	1 $\uparrow$
Acer pseudoplatanus	46	16 $\uparrow\uparrow$	5 $\uparrow\uparrow$
Fagus sylvatica	41	37	14 $\uparrow\uparrow$
Pseudotsuga menziesii	48	44	9 $\uparrow\uparrow$
Abies procera	74	46 $\uparrow$	33 $\uparrow\uparrow$
Fraxinus excelsior	81	35 $\uparrow\uparrow$	2 $\uparrow\uparrow$
Acer platanoides	90	11 $\uparrow\uparrow$	0 $\uparrow\uparrow$

Mean values were used for the rates of loss so far; if the analysis is done spatially, significantly different values were calculated. In 1996 a mean rate of loss of 17% was calculated for the untreated control plot. The base of this calculation, the sub plots, each 225 m^2 in size, with originally 75 spruces, a rate of 7% at best and on the other extreme 38% for the worst case.

Significant differences within the different tree species can be seen comparing the three types of soil cultivation. Spreading 4 t ha^{-1} of magnesian lime, as usually done in liming practice, significantly increased the rate of survival only for noble fir, ash and the two species of maple. The situation for spruce improved significantly as well. This improvement, in comparison to the control plots, was not remarkable though, because this species usually develops well under difficult conditions. The rate of mortality for douglas fir and beech was almost as high as for the reference plots, the pioneer species birch and alder had surprisingly higher mortality rates on the treated plots.

The intensive soil cultivation variation LPF, which is impossible to be realised in practice, had a positive effect on all tree species except for mountain ash. Very demanding species like beech and ash showed a great increase in their rate of survival.

4.2.3.3   Natural reforestation

The wind throw area, 3 ha in size, is fully surrounded by forest. In between the planted trees other trees seeds flew in and regenerated naturally. A representative survey, conducted three years after planting, presented Table 4.2-2.

Table 4.2-2. Natural regeneration three years after stand establishment (Oct. 1993) (Total number per ha and species, relative distribution of species (portion) according to different treatments (Plot) and average height (mean and standard deviation).

Species	Plot	Portion	Height	Data broken down by experimental plots		
		(%)	(cm)	Plot	Species	%
Betula	CON	49	59±24	CON	*Betula*	52
pendula	LIM	33	65±27	Σ 1606 ha⁻¹	*Sorbus*	18
	LPF	18	73±33	broken	*Picea*	27
Σ 1615 ha⁻¹		100		down by	*Salix*	3
Sorbus	CON	61	105±45	LIM	*Betula*	39
aucuparia	LIM	35	71±39	Σ 1491 ha⁻¹	*Sorbus*	11
	LPF	4	53±31	broken	*Picea*	16
Σ 510 ha⁻¹		100		down by	*Salix*	34
Picea	CON	57	17±8	LPF	*Betula*	16
abies	LIM	30	31±20	Σ 2049 ha⁻¹	*Sorbus*	1
	LPF	13	12±4	broken	*Picea*	5
Σ 822 ha⁻¹		100		down by	*Salix*	78
Salix	CON	2	61±19			
caprea	LIM	24	63±24			
	LPF	74	103±46			
Σ 2199 ha⁻¹		100				

The natural regeneration within the culture consisted exclusively of willow (*Salix caprea*), sand birch, spruce and mountain ash. Willow had the highest number with 2200 plants ha^{-1}, mountain ash the smallest with about 500 plants. The succession of seeding density with increasing soil cultivation is conspicuous. Except for the willows all three species grew densest in areas with the worst soil condition. Regardless of the much differing species specific numbers of plants per hectare, half of the recorded regenerated plants of one species grew on the untreated control sub plot CON (49-61%), a third (30-35%) on the LIM sub plot with surface liming and 4-18% on the fully meliorated sub plot LPF (Table 4.2-2). Because the number for willow is the opposite, the total numbers for regenerated plants for all species do not differ too much between the three different sub plots. The species combination is different for the three sub-plots though. It is birch and spruce for the CON plot, birch and willow for the LIM and almost exclusively willow for the fully meliorated LPF variety.

A significant clustering for the tree species, as seen for the loss percentages, was found for the growth parameters as well. As expected, the modest pioneer species mountain ash, birch and black alder proved to be most suitable for the rough site conditions of the untreated soil (control plot CON). Six years after stand establishment these species grew to a height of 2.5 m, some individuals were four meters tall. Black alder showed by far the greatest growth increment compared to its initial height. A significant growth increment was also found for douglas fir and spruce. The height increment was distinctly higher for these two species, after a time of stabilisation between the fourth and sixth year (Fig.4.2-2). Six years after stand establishment beech and noble fir only reached an average growth increment, which corresponds to the initial height of the trees. The valuable broad leafed species developed extremely unsatisfactory: whereas the average height of the remaining trees of the six year old ash culture increased at least a couple of centimetres since planting, the average heights for the two maple species were distinctly smaller (than at the time of planting). This is for one thing an effect of browsing, mainly done by hare. Another important aspect is the mathematical shift, since only every second tree for sycamore maple and every tenth individual for Norway maple were still alive and included in the calculations. Table 4.2-3 shows the mean root collar diameter for the 4-year-old culture, visualising the same trend for the 10 tested species. Species specific differences for slenderness are obvious though. The mean quotient of height and root collar diameter was 120 for spruce and only 70 for ash.

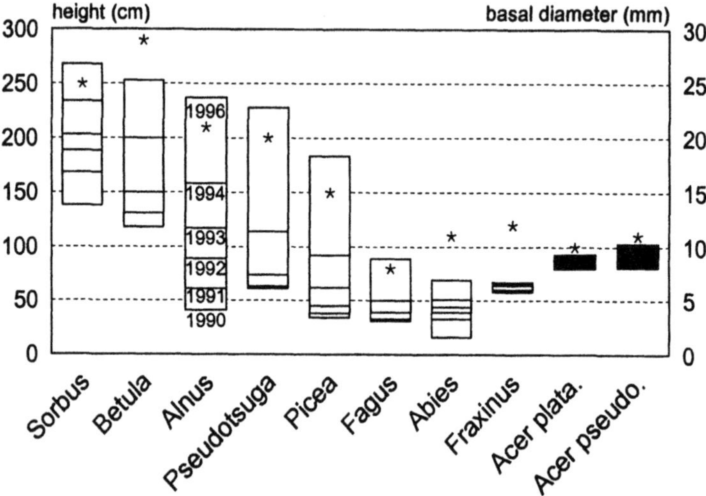

Fig.4.2-2. Height development for the remaining tree species on the untreated control plots in the year of planting (1990) until the end of 1996 (the black columns show a decline of the mean plant height, the asterisks show the root collar diameter (basal diameter) for the 4 year old culture).

Table 4.2-3. Relative comparison of mean tree height and root collar diameter on the research plots (%). The heights of the two, four and six year old cultures of the research variances LIM and LPF are shown, put into proportion to the control area (CON = 100%). The relations for the root collar diameters are only shown for the 4-year-old plants.
↑ tree species that had significant height differences between the three experimental plots immediately after planting. Hence, the presented data are corrected data, the percentages were changed on the base of the plot depending height differences at the beiginning of the experiment.
* significant contrast ($\alpha < 0,05$, analysis of variance ANOVA) as against the control plot (CON)

| Species | Average tree height (%) | | | | | | Diameter (%) | |
| | 2-years (1992) | | 4-years (1994) | | 6-years (1996) | | 4-years (1994) | |
	LIM	LPF	LIM	LPF	LIM	LPF	LIM	LPF
Sorbus ↑	110	102	104	111	88	135*	96	113
Betula	92	115	85*	118*	100	167*	82	112
Alnus ↑	102	123	113	157*	107	151*	113	174*
Pseudotsuga ↑	111	121	118	140*	120	129	104	115
Picea	104	97	110	121*	99	108	98	111
Fagus	103	113	101	182*	100	184*	85	117
Abies	100	99	121	126	130*	150*	110	103
Fraxinus	113	134*	111	189*	122	308*	95	117
Acer platano.	110	127	85	142*	100	234*	130*	164*
Acer pseudo.	115	127	133*	270*	152*	374*	115	206*

Two years after planting a statistically secured increase in height growth (34%) could only be found for ash on the meliorated plot (LPF-variance). The increase in height growth ranged between 10 and 30 for most other species %. Liming with 4t ha^{-1}, as used in practice, lead to a slight increase of increment for the valuable deciduous species only. Compared to the untreated control plot (CON) a significant lead in terms of growth can be recognised for all species of the four-year-old culture, except for mountain ash though.

Mountain maple responded strongly to the improved soil conditions. The moderate liming variance (LIM) only showed a slight improvement compared to the control plot. Two years later the differences for the more demanding species, ash, noble fir and both kinds of maple still increased, comparing the fully treated and the control plot. At the same time the pioneer species showed a distinct increase in height increment compared to the control plot. Spruce showed the least reactions to the more favourable soil conditions (Table 4.2-3). If the same comparison is done for the diameter increment, the varied soil conditions seem to have a minor effect.

The intensive amelioration (LPF) in combination with sowing so called assistant plant cover, lead to a partially dense ground vegetation. This might lead to a competitive situation for the timber species, which might not be compensated by the advantages of an improvement of soil conditions. Considering the very critical first two years after stand establishment in terms of competition, ash and sycamore maple showed a decrease in height when the ground vegetation covered more than 50% of the plot. None of the other species showed statistically proven differences.

4.2.3.4 Nutrient concentrations

Different tree species, showed very differing element contents, under similar site conditions (Table 4.2-4). The differences between the species were more distinct than for the three experimental variations. Spruce, noble fir and douglas fir were poorly supplied with a nitrogen content of 10-12 mg g^{-1}. These tree species have, according to other authors, a nitrogen deficiency at a content of less than 12 mg g^{-1}. These values correspond to very high C/N ratios, ranging between 43 and 48. Slightly higher nitrogen concentrations could be proven for mountain ash and the two maple species. Concentrations of at least 2% were found for beech, black alder and birch. A significant improvement of the nitrogen content due to the amelioration measures could not be statistically proven for any of the species (Table 4.2-5).

The differences for phosphorus content are much smaller, when comparing the different tree species (Table 4.2-4). They range between 1.0 to 1.8 mg g^{-1} dra mass only birch shows up to 3.0 mg g^{-1}. The majority of the samples reached a critical point already, being in correspondence with other authors at 1.5 mg g^{-1} dry mass (Burg 1990). The values for potassium are within a similar narrow scope. Almost all species element contents range between five and seven mg g^{-1} dry mass, only mountain ash and the two maple species showed a higher content with 8 to 11 mg g^{-1}. A satisfying concentration of potassium in the leaves ranges between 11 and 12 mg g^{-1} dry mass for mountain ash (Weihs 1993). For spruce a normal stage of nutrition situation is already reached at 4,5 mg g^{-1} to 8 mg g^{-1} of needle dry mass (Burg 1990).

Calcium, magnesium and manganese belong to a group of elements showing distinct, species specific differences. This is applies markedly to calcium, because the concentration ranges between 2 and 4 mg g^{-1} dry mass for douglas fir, noble fir and spruce and can reach 17 mg g^{-1} for mountain ash. These results are contributed less to the fact, that mountain ash in comparison with the conifers, is better capable of assimilating calcium from the soil. It is much more likely that the annual shifting of calcium from other parts of the plant into the leaves of mountain ash and other deciduous species is almost done, while this process is mainly done in late winter for the conifers. Magnesium, an often insufficiently available element, was found with average concentrations of 0.7 to 1.3 mg g^{-1} for birch. Black alder and noble fir showed values ranging between 1.8 and 2.4 mg g^{-1}. The content of manganese within the assimilation organs for most tree species on the control plot ranged between 3.5 and 6.4 mg g^{-1}, only spruce and douglas fir had values below 2 mg g^{-1}. Therefore the mean magnesium concentrations for the researched tree species differ by the triple, the manganese content by the quadruple. An obvious deficiency for magnesium is not in hand for the magnesium concentrations found here, ranging between 0.7 and 2.4 mg g^{-1}.

Table 4.2-4. Average element concentration for leaves and needles (one year old) for the six year old culture on the untreated control plot CON, September 1996 (all data in mg g^{-1} dry mass, except for the C/N ratio).

Tree species	N	C/N	P	K	Ca	Mg	Mn
Alnus glutinosa	25±2	20	1.4±0.2	6.4±1.1	8.0±1.9	1.8±0.3	4.1±1.7
Sorbus aucupa.	13±2	36	1.5±0.3	8.7±3.4	17.0±2.3	1.3±0.2	5.0±3.4
Betulla pendula	29±4	17	3.0±0.4	7.1±1.4	8.5±2.0	1.8±0.3	6.4±0.2
Picea abies	11±1	48	1.4±0.1	5.7±0.2	3.9±1.6	0.8±0.1	1.4±0.2
Pseudotsuga men.	12±1	43	1.1±0.1	5.3±0.3	2.4±0.8	0.7±0.1	1.6±0.9
Fagus sylvatica	20±2	25	1.8±0.3	5.5±0.7	5.9±1.4	0.8±0.1	5.2±2.1
Acer platanoides	15±2	31	1.0±0.2	8.2±3.2	7.1±1.1	1.3±0.4	5.0±1.8
Acer pseudoplata.	16±4	32	1.1±0.1	11±2.4	9.8±1.4	1.3±0.5	4.7±2.0
Abies procera	11±3	46	1.5±0.4	7.1±1.1	3.4±0.5	2.4±0.7	3.5±1.3

Comparison of nutrient elements for the three treatment varieties lead to the unexpected result, that the only elements showing a significant difference for their mean values compared to the control variety, were magnesium, calcium and manganese (Table 4.2-5). Nitrogen and therefor the C/N ratio, iron and zinc belonged to the group of elements not showing the slightest difference in their content within the assimilation organs of the researched tree species comparing the different treatment varieties. Phosphorus and potassium showed at least for the fully ameliorated plot a significant, but not statistically secured increase in element content. The only elements showing a significantly higher concentration in the above surface parts of the plant, achieved by the soil improvement, are calcium and magnesium. The manganese content shows an opposite trend.

Table 4.2-5. Mean portion of the leaf and needle content (including C/N ratio) for all researched tree species put into relation to the trees on the control plot (CON).

Plot	N	C/N	P	K	Ca	Mg	Mn	Fe	Zn
CON	100	100	100	100	100	100	100	100	100
LIM	97	98	98	100	123	171*	50*	97	108
LPF	101	97	114	113	169*	173*	20*	97	97

*significant contrast (α <0,05, analysis of variance ANOVA) as against the control plot (CON)

The results presented so far were based on the data obtained from the four year old cultures. Analysis of element content for the major tree species beech and spruce was additionally done two years earlier. Comparing the data on element content for these species at the different times shows differences for the species as well as for the treatment varieties (Table 4.2-6).

Table 4.2-6. Element content of spruce and beech on the three differently treated research plots. Sampling was done in 1994 and 1996, concentrations are stated in mg g^{-1} leaf or needle dry mass.

Plot	Tree species	Age (years)	N	P	K	Ca	Mg
					(mg g^{-1} dry mass)		
CON	*Picea abies*	2	11.6	1.5	5.7	8.6	0.5
		4.	10.4	1.4	5.7	3.9	0.8
	Fagus sylvatica	2	14.3	1.2	5.3	7.7	0.6
		4	10.0	1.8	5.5	5.9	0.8
LIM	*Picea abies*	2	13.2	2.1	6.0	9.1	1.3
		4	11.9	1.9	6.8	5.1	0.9
	Fagus sylvatica	2	13.5	1.4	4.5	8.4	1.5
		4	20.8	1.6	4.9	8.2	1.7
LPF	*Picea abies*	2	13.8	2.0	5.9	12.6	1.4
		4	9.8	2.1	8.0	5.8	1.8
	Fagus sylvatica	2	14.7	1.4	5.6	9.2	1.6
		4	18.5	1.5	4.9	11.0	2.2

A slight increase in content for all treatment varieties within the first two years is only found for magnesium. Calcium shows a decrease for all treatment varieties of spruce, the course for beech depends on the treatment: The untreated control plot (CON) shows a modest decrease in concentration, whereas there is no change on the plots with surface liming (LIM). The concentrations on the fully treated plot (LPF) even increase by 20%. This treatment dependent trend applies to both species for potassium. The control plot does not show any changes for the element content of the leaves. The content of nitrogen and phosphorus do not show any kind of trend.

4.2.4 Discussion

The setup of this research shows several deficiencies without any doubt, these had to be accepted due to otherwise the high costs. A statistically dependable coverage on possible differences for the soil treatment could not be achieved, because the treatment was not repeated (Hurlbert 1984). A number of reports state the positive effect of amelioration on problematic sites (Ulrich 1970 and 1991, Gussone 1983, Kenk et al. 1984, Hüttl 1987 and 1990, Zöttl 1990, Klöppel et al. 1991, Kühnhold 1991). Comparative studies on the impact of liming and fertilisation are rare though (Lezer et al. 1979, Weihs 1993).

The loss of nutrients on forest sites cannot only be connected to the acidification of soils, the increase in nitrogen input also leads to an impoverishment in magnesium and calcium (Ulrich 1975). Fertilizers containing magnesium do not only lead to an improvement in soil condition, they might also have a significant influence on reproduction (Wenzel 1989, Katzensteiner et al. 1995, Jandl 1996). This type of treatment often causes the leaching of nitrate, which is limited to a couple of years though (Wenzel 1989, Katzensteiner et al. 1995).

Fertilisation with calcium and phosphorus lead to an increase in growth of 11% for spruce on acidified similigley, a further addition of nitrogen and potassium lead to an increase of 20%. Nutrient analysis proved that the growth increment of spruce improved, with an increasing content of nitrogen, phosphorus, potassium and magnesium (Hunger 1979). It has to be put into consideration though, that the element content of the needles fluctuates within a great range annually. The hydrology has an important impact as well: Stagnant moisture obstructs the assimilation of phosphorus, nitrogen, potassium and manganese.

Research conducted on acidified forest sites in the Ardenne mountains showed, that spruce cultures treated with NPP-fertiliser had a higher rate of survival due to an improvement of their nutrient situation. Furthermore, growth increased 20-30%. Soil cultivation without fertilisation lead to a significant increase compared to the untreated control plot. Positive effects of fertilisation could be furthermore proven for the ground vegetation, soil-micro fauna and root distribution (Weissen 1977).

As shown under similar site conditions, an improvement in calcium supply has definite advantages for tree growth, as well as for the floristic variety and the humus condition (Fehlen and Picard 1994). The concern, that fertilising may be beneficial to the associated flora and therefore a threat for the timber species, does not seem to be justified. It is true, the associated flora develops very well after soil cultivation but this does usually not endanger the timber species due to their delayed promotion (Garbaye and Bonneau 1975).

The unsatisfying results for beech on the untreated control plot are less explained by the rough climatic conditions but by the soil condition. It cannot be totally clarified, if this is a question of a high sensitivity towards aluminium-ions within the soil solution (Neitzke and Runge 1987) or a high concentration of protons (Ljungstrom and Stjernquist 1993).

It is not contentious that a poor supply of nutrients in combination with sulphur dioxide increases the strain on the trees (Jurat et al. 1986). Soil cultivation (amelioration) leads to a considerable increase in the rate of survival and growth. An indirect optimising in resistance against meteorological risks cannot be excluded: Senser et al. (1990) found an increase in frost sensitivity for young spruce with a calcium- and magnesium deficiency on acidic sites. Lepoutre and Cros proved in 1979, that origin of the timber species is not only important in regards to climatic adaptation but also in terms of adaptation to soil conditions: Seeds originating from lime sites developed best on highly alkaline soils (pH 8), while young plants originating from beech forests on acid soils, grew well on acidic soils (pH 4.9). Leaf analysis showed, that the calcium assimilation of the acidic origins was much too high and lead to relative lack of phosphorus and manganese. This is not expected at the grade of acidification of the soils in the presented research project. Weihs (1993) conducted a four year project in the Harz mountains. He combined species and fertilisation experiments on different problematic sites. The results are well comparable to the research presented in this study. The ranking of mortality percentages on untreated soils is similar to the ranking found in this study, the differences between the tree species are smaller though. After four growing periods and with hardly any crown coverage (crown coverage from remaining old growth <25%) mortality for mountain ash was 10%, 18% for black alder, 32% for spruce, 45% for sycamore maple and 55% for *Abies procera*. The provenance chosen for white alder (*Alnus incana*) was not adapted to the site conditions, which lead to an unexpectedly high mortality rate of 54%. Pacific fir is very sensitive to replanting, the "planting shock" usually lasts for several years (Feinauer 1991) and leads to the bad results in the described case as well as in the conducted study. The cultivation measures, plant hole liming combined with amelioration, split the losses for noble fir, mountain ash and white alder in half and reduced them to a third for spruce. This as well showed, in correspondence with the results presented in this study, an improvement for sycamore maple. Black alder did not show any reaction on these supportive treatments.

4.2.5 Conclusions

The sylvicultural opportunities to maintain the stability of our forest ecosystems are limited due to severe environmental pollution. Single, positive reports stating, that early thinning and promoting dominant, well shaped trees leads to a significant improvement in terms of vitality, can hardly change this. The choice of suitable tree species, that are able to develop satisfactory under difficult circumstances, present a central, silvicultural goal. The observation over several years presented in this study, shows considerable differences between the different species. A stronger emphasis has to be put on pioneer species, to avoid an extremely costly treatment for soils on problematic sites in the future. Spruce will, under this aspect, also play an important role in silvicultural planning.

Acknowledgements

Amoung the many persons who contributed to research at the amelioration experiment and deserve acknowledgement, I want to give my special thanks to the biologist T. Gieger as well as the technicians D. Pryor, G. Brauer, A. Gerke, A. Nannen, C. Suner, K.-H. Obal, F. Schüddekopf, K.-H. Heine and U. Schmidt. The study received finacial support from the German Federal Ministry of Research and Technology and the state of Lower Saxony.

References

Binder F (1991) Aufforstung in Waldschadensgebieten. Forstl Forschungsberichte München 119

Burg J v d (1990) Foliar analysis for determination of tree nutrient status a compilation of literature data. In: De Dorschkamp, Institute for Forestry and Urban Ecology Wageningen, Niederlande, Rapport Nr 591

Fehlen N, Picard JF (1994) Influence de la fertilisation sur la végétation spontanée et la croissance radiale de l'épicea commun (*Picea abies*) - The effect of fertilizer on natural vegetation and radial growth of Norway spruce (*Picea abies*) in an adult platation. Annales Sciences Forestières 51:569-580

Feinauer H (1991). Pazifische Edeltannen im Hochschwarzwald. Allg Forst Jagdztg 46:237-238

Garbaye J, Bonneau M (1975) Prelimary results from a fertilizer trial on *Quercus sessiliflora*. Annales Sciences Forestières 32:173-183

Gravenhorst G, Szarejko Z (1990) Das Klima des Sollings (in: Exkursionsführer Solling - Oktober 1990). Ber Forschungszentrum Waldökosysteme, Göttingen, Reihe B, Bd 17:5-14

Gussone HA (1983) Die Praxis der Kalkung im Walde der Bundesrepublik Deutschland. Forst Holz 38:63-71

Hunger W (1979) Ernährung und Düngung der Fichte auf Pseudogleystandorten des Sächsischen Hügellandes. Wiss Zeitschr Techn Univ Dresden 28:1349-1359

Hurlbert SH (1984) Pseudoreplication and the Design of Ecological Field Experiments. Ecological Monographs 54:187-211

Hüttl RF (1987) "Neuartige" Waldschäden, Ernährungsstörungen und Düngung. Allg Forstz 42:289-299

Hüttl RF (1990) Vergleichende Analyse von Düngungsversuchen in der Bundesrepublik Deutschland und in den USA. Forst Holz 45:580-587

Jandl R (1996) Magnesiumdüngungsexperiment Sulzberg (Magnesium fertilizer experiment at Sulzberg) Centralblatt Gesamte Forstwesen 113:131-154

Jurat R, Schaub H, Stienen H, Bauch J (1986) Einfluß von Schwefeldioxid auf Fichten (*Picea abies* [L.] Karst.) in verschiedenen Bodensubstraten. Forstwiss Cbl 105:105-115

Katzensteiner K, Eckmüllner O, Jandl R, Glatzel G, Sterba H, Wessely A, Hüttl RF, Nilsson LO (ed) (1995) Revitalization experiments in magnesium deficient Norway spruce stands in Austria. Nutrient uptake and cycling in forest ecosystems. Plant and Soil 168-169:489-500

Kenk G, Unfried P, Eevers FH, Hildebrand EE (1984) Die Düngung zur Minderung der neuartigen Waldschäden - Auswertungen eines alten Düngungsversuchs zu Fichte im Buntsandstein-Odenwald. Forstw Cbl 103:307-320

Klöppel H, Kördel W, Fabig W (1991) Auswirkungen unterschiedlicher Kalkungsvarianten im Sauerland. Allg Forstz 46:67-69

Kühnhold K (1991) Reaktionen einer Buchen-Douglasien-Kultur auf verschiedene Düngungs- und Meliorationsmaßnahmen in der Wingst. Diplomarbeit Univ Göttingen

Lepoutre B, Cros ET (1979) Growth and nutrition of one-year-old beech (*Fagus sylvatica* L.) seedlings of different provenaces on a natural soil or the same soil with added lime. Annales Sciences Forestières 36:239-262

Ljungstrom M, Stjernquist I (1993) Factors toxic to beech (*Fagus sylvatica L.*) seedlings in acid soils. Plant and soil 157:19-29

Meiwes KJ (1994) Erstellung einer Versuchsanlage und Stoffhaushaltsuntersuchungen im Projekt Meliorationsversuch mit Mischbestandsbegründung im Solling. Abschlußbericht 1989-1993 zum BMFT Forschungsvorhaben „Stabilitätsbedingungen von Waldökosystemen. Ber Forschungszentrum Waldökosysteme, Göttingen, Reihe B, Bd 37:508-515

Melzer EW, Lucke E, Hertel HJ (1979) Bedeutung von Düngung, Melioration und Baumartenmischung für die Entwicklung der Fichte in Kultur und Jungbestand. Wiss Zeitschr der TU Dresden 28:1319-1323

Neitzke M, Runge M (1987) Entwicklung und Mineralstoffgehalte junger Buchen in Abhängigkeit von den Aluminium- und Kalziumgehalten der Nährlösung. Teil 1: Entwicklung. Bot Jahrbuch Syst 108:403-495

Senser M, Blank LW (eds), Lutz C (1990) Influence of soil substrate and ozone plus acid mist on the frost resistance of young Norway spruce. Tree Exposure Experimental Pollution 64:265-278

Ulrich B (1970) Die Reaktion von Calciumcarbonat bei der Einarbeitung von Kalkmergel in stark versauerten Waldböden mit Auflagehumus. Allg Forst Jagdztg 141:5-9

Ulrich B (1975) Die Umweltbeeinflussung des Nährstoffhaushaltes eines bodensauren Buchenwalds (Enviromental influences on the nutrient regime of a beech forest on acid soil) Forstw Cbl 94:280-287

Ulrich B (1991) Folgerungen aus 10 Jahren Waldökosystem- und Waldschadensforschung. Forst Holz 46:575-588

Weber M (1991) Waldbauliche Untersuchungen zu den neuartigen Waldschäden in jungen Fichtenbeständen Ostbayerns. Forstliche Forschungsberichte München 109

Weihs U (1993) Walderneuerung auf Problemstandorten der Harzhochlagen - ein Baumartenanbau- und Düngeversuch. Ber Forschungszentrum Waldökosysteme, Göttingen, Reihe A, Bd 102

Weissen F (1977) Dix années d'évolution d'une deuxième generation d'epiceas fertilisés à la plantation (Ten years development of second generation spruce stand fertilized at planting). Bulletin Recherches Agronomiques Gembloux 12:1-2, 135-157, 159-171

Wenzel B (1989) Kalkungs- und Meliorationsexperimente im Solling: Initialeffekte auf Boden, Sickerwasser und Vegetation. Ber Forschungszentrum Waldökosysteme, Göttingen, Reihe A, Bd 51

Zöttl HW (1990) Ernährung und Düngung der Fichte. Forstw Cbl 109:130-137

4.3 Agricultural herbs as auxiliary plants in stand establishment

A. Roloff, J. Linke

Institute of Forest Botany and Forest Zoology, Dresden University of Technology,
Pienner Straße 7, D-01735 Tharandt, Germany,
e-mail: forstbot@forst.tu-dresden.de

Abstract

The article focuses on the assessment of an experiment establishing an amelioration plot on the site of a former mature spruce stand. In this connection various treatments were applied: deep ploughing including liming/fertilisation-liming, superficial liming/fertilisation-liming, and control. Sowing agricultural auxiliary plants each took place on half the area of the treated variants with the objective to accumulate a possibly big amount of biomass during a short period in order to cover the soil in this way and to integrate the applied nutrient elements into the element cycle. The colonisation dynamics, the development of the biomass as well as the elements stored in the agricultural auxiliary plants are examined and presented, aided by so-called permanent monitoring plots or by harvesting the biomass in trial plots, respectively. Based on the data obtained, the application of agricultural auxiliary plants on cleared sites which are provided for forest cultivation is appraised.

Key words: agricultural auxiliary plants, afforestation, dynamics of plant colonisation, suppression of a grass stage, biomass, integration of nutrient elements, distortion of the flora

4.3.1 Introduction

In 1989 the experiment "Ecosystem amelioration by liming/fertilisation-liming, tillage and cultivation of auxiliary plants" was drawn up within the framework of investigations into stability conditions of forest ecosystems (Ulrich 1989).

In addition to the nutrient-analytical investigations of the species of the ground vegetation as well as vegetation-ecological investigations during the initial phase of the secondary succession (Linke 1994), a main point of the experiment was the use of auxiliary plants in stand establishment. The latter is to be dealt with in closer detail in the following.

The seeding of agricultural auxiliary plants took place under the following aspects:

1. the species were supposed to form a lot of biomass during a short time period:
    - for reasons of erosion protection, especially on the deep ploughing variant, and
    - for the purpose of integrating the artificially applied nutrients into the element cycle via the biomass, thus supplying the forest crop plants with so-called initial capital;
2. a long-term grass stage, as becomes evident on the control plot (with *Avenella flexuosa, Agrostis tenuis* etc.), has to be prevented by the agricultural species.

Other aims pursued by seeding are to control secondary succession and to decrease the hazard of feeding by mice on crop tree species, since grass is a biotope preferred by mice.

The mixture of agricultural auxiliary plants was developed at the Forest Research Institute of Lower Saxony (young growth tending and site rehabilitation by crop plant cover-field trial: Reinecke 1990) and adjusted to the site conditions of the trial plot (Reinecke 1990). It comprises the following species:

Trifolium hybridum	alsike clover (+ Radzin) HYTRA
Trifolium incarnatum	crimson clover
Trifolium subterraneum	subterranean clover (+ Radizin) HYTRA
Phacelia tanacetifolia	Phacelia, honey flora (honey-producing plants)
Raphanus sativus	oil radish (winter variety)
Brassica rapa spp. *oleifera*	wild turnip NOKO
Malva sylvestris	mallow
Sinapis alba	white (yellow) mustard

Subterranean clover and alsike clover were infected with Radizin bacterial cultures. The three clover species are not further differentiated in the investigation, but merely referred to as "clover".

4.3.2 Forest community and trial plots

4.3.2.1   *Potentially natural and actual forest community*

The potentially natural *Luzulo-Fagetum* in the Solling was largely replaced by spruce forests centuries ago (Gerlach et al. 1970). The most frequent species of the typical spruce forest, in which a thick layer of litter has accumulated, are several mosses growing on raw humus such as *Dicranella heteromalla* and *Polytrichum formosum*, in addition *Vaccinium myrtillus, Galium harcynicum, Luzula luzuloides, Avenella flexuosa, Carex pilulifera,* and in more light-permeable stands and in stand gaps *Epilobium angustifolium, Rumex acetosella, Rubus idaeus, Digitalis purpurea*, partly also *Trientalis europaea*. The latter is the differentiating species of an own community, the star flower (*Trientalis*)-spruce forest which is an indicator of damp, often deep soils distinguished by a higher proportion of loess. This community is greatly distributed especially in plateau positions.

4.3.2.2   Trial plots

The trial area is situated in the Solling mountains in plateau position at about 500 m a. s. l. The site developed from a loess layer (50 cm to 60 cm in thickness) and an underlying Triassic red sandstone. It is heavily acidified and has found to be within the aluminum buffer range (for details see Meiwes 2002, this volume).

The average annual mean temperature over many years is 6.4 °C, and that of the growing season (May to September) 12.3 °C. The annual mean of precipitation accounts for 1080 mm (Torfhaus measuring station). The mature spruce stand stocking there partly declined due to wind-throw in autumn 1987. The remaining stand was felled in winter 1989, and in June 1990 the tree-stump removal by an excavator took place. The individual variants are delimited by felling refuse (slash) and stumps which were pushed together by a bulldozer and a heavy-duty cultivator to form embankments.

The three variants dealt with consist of deep ploughing, superficial liming/fertilisation-liming, and a control; each of the three variants comprised about one hectare. The experimental design is described in detail by Meiwes (2002, this volume)

On the plot of the deep ploughing variant 18.5 t of magnesium-containing lime per hectare were mixed into the soil up to a depth of 70 cm, 4.5 t of dolomite were superficially intermixed using a disk harrow. The fertilisation which was performed together with dolomite liming consisted of: 50 kg P ha^{-1} (in the form of superphosphate 120 kg P_2O_5 ha^{-1}) and 50 kg K ha^{-1} (in the form of potassium sulphate).

In the variant of the superficial fertilisation the liming consisted of 1.5 t ha^{-1} of finely ground chalk, and the fertilisation of 500 kg ha^{-1} dolomite and 50 kg P ha^{-1}.

The crop tree species were planted in all variants within squares of 15 m x 15 m, in alternating order in one square beech trees (*Fagus sylvatica*, 50 to 70 cm, 2 + 0) in a spacing of 1.5 m x 1 m^2, in the next square spruce trees (*Picea abies*, 50 cm to 70 cm) in a spacing of 3 m x 1 m^2, and across all the squares black alders (*Alnus glutinosa*, 120 cm to 150 cm in height) in a spacing of 4 m x 2.5 m^2.

4.3.3 Methods

The vegetation surveys performed in so-called permanent monitoring plots (1 DBL = 10 x 1 m^2 subplots) also cover the sown agricultural species in addition to the site-related, i.e. typical vegetation indicative of the site relations, thus giving an insight into the colonisation dynamics on the differently treated plots. Estimation of vegetation coverage by samples was carried out twice a year (6.91: June 1991, 9.91: September 1991, 6.92: June 1992 and 9.92: September 1992) by estimating the species dominance and the overall dominance in each of the ten subplots of a permanent monitoring plot. Estimation of the dominance is based on the scale of Londo (1975). Each four permanent monitoring plots are included in the survey of the deep ploughing and of the superficial fertilisation-liming variants, while in the control variant there are six plots (see Linke 1994).

The biomass was determined at intervals of two months – at three dates in the growing season from May to October (6.91, 8.91, 10.91, 5.92, 7.92, 9.92). Per variant 12 subplots of 1 m² were harvested. The biomass samples which were harvested at those dates were used for the element analyses.

Each on half of the deep ploughing and half of the superficial fertilisation-liming variant a mixture of agricultural herbs was sown in May of 1991, which emerged quite acceptably (Linke 1994).

4.3.4 Results

4.3.4.1   *Dynamic of colonisation*

The number of species growing on the deep ploughing variant including agricultural auxiliary plants shows (Table 4.3-1) that the agricultural herbs, which already experienced a strong distribution during the first year, tend to impair the invasion of other species. Only in the second year the species number exceeded that of the deep ploughing variant without agricultural auxiliary plants (see Linke 1994), with two of the sown species - viz phacelia and mallow - being no longer present.

The high absolute frequency of the clover species (36/39/40/40) which occur on all subplots is especially obvious in the second year. At the same time, *Agrostis tenuis, Galium harcynicum, Epilobium angustifolium* and oil radish as the second agricultural species are very frequently recorded on the plots (32/40/29/26).

Like *Agrostis tenuis, Avenella flexuosa* is regarded separately from the group of 'graminales' because of its pronounced presence and colonisation dynamics over the whole plot. It accounts, though likewise being represented in high frequency (8/9/23/21), of less than 10% of the overall dominance. It becomes, however, obvious that the sown auxiliary plants highly dominate at the beginning of the investigation (6.91 and 9.91); in contrast to this *Agrostis tenuis* occupies an ever increasing area (relat. dominance at the date 9.92 = about 30%; frequency = 22/29/34/37). Except for the species named before, only the moss species (about 15%) reach a proportion of more than 10% in the overall dominance (see Linke 1994).

The vegetation surveys conducted in the four permanent monitoring plots (PMP) regarding the variant of superficial fertilisation-liming which includes the agricultural auxiliary plants lead to the following result (Table 4.3-2).

Table 4.3-1. Species inventory (frequence) on the ploughland-version including auxiliary agricultural plants.

| Species | Frequency regarding all permanent monitoring plots | | | |
| | Dates of survey (month.year) | | | |
	6.91	9.91	6.92	9.92
Agrostis tenuis	22	29	34	37
Avenella flexuosa	8	9	23	21
Epilobium montanum	3	0	12	14
Epilobium angustifolium	0	5	23	23
Galium harcynicum	7	19	28	29
Carex pilulifera	0	10	12	14
Carex pallescens	0	0	1	0
Carex sylvatica	0	0	0	1
Luzula luzuloides	0	1	1	1
Rubus fruticosus	0	0	1	3
Rubus idaeus	0	0	1	0
Taraxacum officinale	0	0	1	1
Juncus effusus	0	1	4	6
Poa annua	1	4	1	1
Calamagrostis epigejos	0	0	1	1
Agropyron repens	0	1	3	3
Moehringia trinervia	0	0	2	0
Phleum pratense	0	0	2	3
Festuca rubra	0	0	2	1
Senecio vulgaris	0	0	3	2
Senecio sylvaticus	0	0	3	1
Tussilago farfara	0	0	1	1
Oil radish	32	40	29	26
Wild turnip	6	37	19	8
White (yellow) mustard	22	10	10	5
Mallow	23	16	0	0
Phacelia tanacet.	10	14	0	0
Clover	36	39	40	40
Dicranella heteromalla	0	0	10	17
Ceratodon purpurens	0	9	33	40
Polytrichum formosum	0	5	26	20
Pohlia nutans	0	7	0	0
Picea abies	0	2	2	4
Salix sp.	0	1	0	0
Alnus sp.	0	1	0	0
Number of species	11	21	30	27

Table 4.3-2. Species inventory (frequence) on the superficial fertilizingliming-version including agricultural auxiliary plants.

Species	Frequency regarding all permanent monitoring plots			
	Dates of survey (month.year)			
	6.91	9.91	6.92	9.92
Agrostis tenuis	21	27	27	29
Avenella flexuosa	32	33	34	36
Epilobium montanum	12	10	5	13
Epilobium angustifolium	10	19	26	29
Galium harcynicum	32	36	38	36
Galium aparine	0	0	1	0
Carex pilulifera	22	28	21	29
Carex pallescens	0	9	6	3
Carex leporina	0	3	8	10
Holcus mollis	1	2	3	2
Luzula luzuloides	0	1	1	3
Luzula campestris	0	0	3	0
Rubus fruticosus	2	1	4	8
Rubus idaeus	0	4	0	0
Stellaria media	2	4	0	0
Cardamine pratensis	1	1	0	0
Vaccinium myrtillus	3	2	3	2
Trientalis europaea	9	8	11	9
Taraxacum officinale	0	1	0	0
Juncus effusus	1	16	16	23
Poa annua	2	2	0	0
Dryopteris carthusiana	0	1	0	0
Rumex acetosella	0	0	2	0
Rumex crispus	0	2	1	1
Urtica dioica	0	0	0	1
Leontodon autumnalis	0	0	1	2
Moehringia trinervia	4	1	0	0
Festuca rubra	0	0	2	0
Senecio vulgaris	0	0	2	1
Polygonum bistorta	0	0	0	1
Calluna vulgaris	0	0	1	0
Oil radish	30	36	12	1
Wild turnip	30	38	1	0
White (yellow) mustard	30	15	0	0
Mallow	30	3	0	0
Phacelia tanacetifolia	21	17	0	0
Clover	13	26	30	30
Dicranella heteromalla	0	4	2	8
Ceratodon purpurens	0	1	2	13
Polytrichum formosum	0	2	0	7
Pohlia nutans	0	2	2	2
Salix sp.	1	2	5	4
Number of species	22	33	29	26

In this variant the species of the highest frequency are *Galium harcynicum* (32/36/38/36) and *Avenella flexuosa* (32/33/34/36), followed by *Carex pilulifera* (22/28/21/29), *Agrostis tenuis* (21/27/27/29) and *Epilobium angustifolium* (10/19/26/29). Compared to all named species the latter propagates most strongly. Except for *Epilobium angustifolium*, all those species had already been on the plot as integral parts of the spruce forest community (Table 4.3-3) and therefore had, already, at the first date of the survey (6.91), a high absolute frequency. Except for the clover species, the sown agricultural auxiliary plants spread very rapidly on this variant; in the second year, however, they had virtually disappeared. In contrast to this, coverage of the clover species increased, and they were represented in 75% of the parcels at the surveys 6.92 and 9.92 (frequency 13/26/30/39).

Table 4.3-3. Species inventory (frequence) of the control.

| Species | Frequency regarding all permanent monitoring plots | | | |
| | Dates of survey (month.year) | | | |
	6.91	9.91	6.92	9.92
Agrostis tenuis	7	9	8	12
Avenella flexuosa	56	56	59	60
Epilobium angustifolium	3	3	2	2
Galium harcynicum	43	49	48	51
Carex pilulifera	36	44	37	49
Carex echinata	0	0	2	0
Carex leporina	0	0	11	10
Luzula luzuloides	1	0	7	8
Luzula campestris	0	3	5	6
Rubus fruticosus	1	0	2	1
Rubus idaeus	0	2	0	0
Oxalis acetosella	0	1	0	1
Vaccinium myrtillus	16	23	21	23
Trientalis europaea	20	17	24	21
Juncus effusus	0	11	17	16
Juncus squarrosus	7	5	0	0
Poa annua	2	0	0	0
Dryopteris carthusiana	4	5	1	5
Athyrium filix-femina	0	0	4	0
Rumex crispus	1	1	1	1
Calluna vulgaris	1	5	12	22
Dicranella heteromalla	21	34	9	23
Ceratodon purpurens	2	5	11	4
Polytrichum formosum	0	20	13	21
Pohlia nutans	0	8	3	5
Picea abies	0	0	0	3
Betula sp.	0	2	2	2
Number of species	16	20	22	22

Only one species, viz *Avenella flexuosa*, has a particularly high relative dominance at the date of the first survey 6.91 (ca. 45%). At the 2nd date (9.91) the distribution of relative dominance is already more balanced in this variant. The pattern is determined by six species or groups of species, respectively. Along with the decreasing proportion of the agricultural auxiliary plants (12.7/16.2/7.4/4.0) mainly the relative dominance of *Avenella flexuosa* increased again (Linke 1994).

4.3.4.2   *Biomass*

The agricultural species on the deep ploughing variant (including the agricultural auxiliary plants) cause the proportion of graminales (except for *Avenella flexuosa* and *Agrostis tenuis*) which join the biomass to remain relatively low even after two years (ca. 5%). *Agrostis tenuis* achieves the highest proportion among the biomass with about 53% (9.92), the group of herbaceous plants (except for *Galium harcynicum*) accounts for about 13% (9.92). The length of time during which this stage will continue is decisively dependent on the growing behaviour of the crop trees (degree of canopy closure – radiation relations/light intensity).

Table 4.3-4a-c. Participation (%) of certain species or groups of species at the medium above-ground total-biomass on the ploughland-version including agricultural auxiliary plants (4a), superficial fertilizing version including agricultural auxiliary plants (4b) and control (4c).1
6.91 = date (June) of harvest in 1991; number behind / = quantity of species within species group; differences to 100 % caused by rounding up / down

Table 4.3-4a

Species and species groups	Dates of harvest					
	6.91	8.91	10.91	5.92	7.92	9.92
Avenella flexuosa	0.8	2.1	0.0	1.8	2.0	0.2
Agrostis tenuis	99.2	17.0	30.4	44.4	40.8	52.6
Galium harcynicum	0.0	0.1	0.6	1.3	7.3	2.0
Gramineous plants	0.0	0.4/1	0.8/3	2.0/5	1.0/5	5.1/4
Herbaceous plants	0.0	0.7/4	0.03/1	2.5/5	5.4/9	12.9/6
Pioneer trees	0.0	0.0	0.1/1	0.0	1.1/2	0.0
Ferns	0.0	0.0	0.0	0.0	0.0	0.0
Mosses	0.0	0.0	0.0	0.0	1.0/1	4.2/3
Agric. auxil. plants	0.0	79.7/8	68.1/6	48.0/3	41.4/3	23.1/3
Total (kg ha^{-1})	8.1	440.4	2076.9	1288.7	2987.7	2899.7

Table 4.3-4b

Species and species groups	Dates of harvest					
	6.91	8.91	10.91	5.92	7.92	9.92
Avenella flexuosa	63.0	67.3	21.4	45.7	16.7	33.2
Agrostis tenuis	31.6	17.6	20.2	10.6	37.9	17.3
Galium harcynicum	3.4	5.9	21.5	15.4	8.8	12.4
Gramineous plants	0.5/2	3.8/4	16.1/6	13.7/3	25.8/6	31.2/6
Herbaceous plants	1.5/5	6.2/8	13.5/7	10.8/7	4.5/11	4.6/7
Pioneer trees	0.0	0.0	4.7/2	0.0	0.01/2	0.0
Ferns	0.01/1	0.2/1	0.0	0.0	0.0	0.0
Mosses	0.0	0.0	0.0	0.4/1	0.4/2	0.8/2
Agric. auxil. plants	0.0	1.2/6	2.6/3	3.4/3	5.8/2	0.6/1
Total (kg ha^{-1})	1028.7	2530.4	1700.5	2736.5	4183.3	3104.8

Table 4.3-4c

Species and species groups	Dates of harvest					
	6.91	8.91	10.91	5.92	7.92	9.92
Avenella flexuosa	95.2	81.8	41.4	52.8	61.6	42.3
Agrostis tenuis	..0.0	..0.5	4.0	0.0	0.1	26.5
Galium harcynicum	1.7	2.0	26.4	25.1	17.6	21.3
Gramineous plants	0.4/2	9.1/2	26.3/5	12.7/2	10.4/3	4.8/3
Herbaceous plants	2.7/4	2.1/2	2.0/3	8.2/5	7.5/4	4.3/4
Pioneer trees	0.0	0.0	0.0	0.0	0.0	0.6/1
Ferns	0.1/1	..0.0	0.0	0.0	0.0	0.0
Mosses	0.03/1	1.5/2	0.0	1.2/1	0.6/2	0.2/1
Agric. auxil. plants	0.0	0.0	0.0	0.0	0.0	0.0
Total (kg ha^{-1})	1251.8	1878.3	893.6	1708.6	22224.4	2196.1

In the superficial fertilisation-liming variant, which comprises agricultural auxiliary plants, the second dates of harvest (8.91/7.92) each give the highest aboveground biomass (Table 4.3-4b). When the two variants are compared, the biomass of the agricultural species in the superficial fertilisation-liming variant only accounts for a very low proportion (0.1/1.2/2.6/3.4/5.8/0.6), whereas, besides *Agrostis tenuis* (99.2/17.0/30.4/44.4/40.8/52.6), the agricultural species appear to have the highest proportion of biomass (0.0/79.7/68.1/48.0/41.4/23.1) in the deep ploughing variant. Although the absolute above-ground total biomass in this variant approaches nearly 3000 kg ha^{-1} in the second year (7.92, 9.92), it is still below that of the superficial fertilisation-liming variant.

A pronounced and obviously enduring grass stage as it is also described by Pfalz et al. (1991) seems to develop in the "undisturbed" control variant, where the above-ground total biomass is distinctly below that of the treated variants (Table 4.3-4c). *Agrostis tenuis, Avenella flexuosa* and the group of graminales altogether account for 73.6% of the total biomass. But also on the superficially fertilised liming variant which includes agricultural auxiliary plants a similar development becomes obvious, although the 'herbaceous plants' are distinctly greater in number. *Avenella flexuosa, Agrostis tenuis* and the group of 'graminales' give a relative dominance of 82% of the above-ground biomass.

In general, the species and biomass composition is determined by the influence of the agricultural auxiliary plants during the first year only. In the second year the short-lived plant species have almost vanished.

The distribution of the dominance/degree of coverage (Linke 1994) and the biomass proportions show that in the deep ploughing variant mainly *Agrostis tenuis* and the group of graminales are able to compete successfully against the agricultural species. Herbaceous plants occupy only a small proportion. The auxiliary plants sown on the superficial fertilisation-liming variant are obviously yet less influential. Concerning a control of the secondary succession, especially the suppression of a long-lasting grass stage this phenomenon proves that it is worthwhile to use site-related forest herbs such as *Rumex acetosella, Taraxacum officinale, Senecio sylvaticus, Moehringia trinervia, Cardamine flexuosa, Stellaria media* and others instead of short-lived agricultural species which do not belong to the native flora. Even if the biomass proportion of the group of graminales and *Agrostis tenuis* is still comparatively low, there is no chance for site-related herbs to establish in addition to these groups of species because of the competition of agricultural species. On the contrary, the agricultural auxiliary plants are an additional factor of competition regarding the establishment of the (desired) site-related herbs during this period.

4.3.4.3   *Elements stored in the agricultural auxiliary plants*

The element analysis of agricultural herbs is subsequently differentiated according to the nitrogen-fixing clover species (mixture of 3 clover species) and a mixture consisting of all other species sown. These other species are compared to the clover species as 'agricultural herbs'. At two dates of the survey there is either no sufficient plant material available for element analysis (6.91) or 'agricultural herbs' do no longer occur in the sampling subplots (9.92) except for the clover species.

In this connection it is interesting to know if there are (in comparison with the naturally growing species) additional elements fixed in the sown agricultural auxiliary plants and thus integrated in the element cycle. For this reason the development of the most important elements stored in agricultural auxiliary plants is presented in the following.

The amount of elements stored is given in terms of summations for each growing season. In this way the actual, absolute storage at the beginning of a year (6.91 and 5.92) to which the subsequently stored amounts are added is apparent. The rate of production can be estimated by means of the 'increments' of the columns or by the difference of the sums of stored substances.

Besides the element contents the biomass is a crucial factor in the calculation of the stored amounts which are summarised as mean values in Table 4.3-5.

As an example of all elements investigated (s. Linke 1994), the development of N, Ca, Mg and Al storage is presented (Fig. 4.3-1 and 4.3-2). Table 4.3-5 and Figures 4.3-1 and 4.3-2 show that the development of the amounts stored clearly depends on that of the biomass. In the deep ploughing variant which includes agricultural auxiliary plants (Var. 1b) the overall quantity of stored elements is strongly influenced by the enormous increase of biomass of 'agricultural herbs' at the 10.91 date of survey (from altogether 257 kg ha^{-1} on 1272 kg ha^{-1}) as well as by the increase of biomass of the clover species at the 7.92 date of the survey (from 426 kg ha^{-1} to 1200 kg ha^{-1}).

In the superficial fertilisation-liming variant with agricultural auxiliary plants (Var. 2b), in which the biomass of the 'agricultural herbs' largely lags behind that of the deep ploughing variant (Var. 1b), comparatively low changes in the stored quantities become apparent. It is merely the increase of clover species biomass at the 2.92 date of the survey regarding N, Ca and Mg which becomes obvious in this variant.

Concerning the stored Al-mounts there are large differences between these two variants. While the clover species in the deep ploughing variant (Var. 1b) accumulate very high Al-quantities during both growing seasons, the remaining 'agricultural herbs' fix very high Al-amounts in the first year only – a fact which is attributed to the distinctly lower biomass during the 2nd year. Partly comparable to the biomass values on the superficial fertilisation-liming variant, the amounts stored here are by far not as large – which points to the fact that the mean aluminum content in the deep ploughing variant is higher. Regarding the question for the cause of the increased supply of plant-available aluminum and the higher rate of uptake in this variant, it is also noteworthy that the forest floor which was ploughed under and which contained aluminum in organically fixed complexes is apparently very important. In this context, reference is made to Linke (1994). Here it is noteworthy that the ground vegetation of the deep ploughing variant obviously functions as a sink for aluminum which otherwise would reach the groundwater via the percolating water and impose considerable loads (also see Schieri et al. 1986, Wenzel 1989, Hildebrand 1990).

In addition the relatively high sum of nitrogen stored in the clover species is remarkable. Despite the lower amount of biomass of clover species in the deep ploughing variant (Var. 1b), during the first year those species fix distinctly greater amounts than the other species. Here the mutualism of the clover species and the nitrogen-fixing nodule-bacteria is obviously without any effect.

Table 4.3-5. Medium above-ground biomass of agricultural auxiliary plants.
Var. 1b = ploughland variant + agricultural auxiliary plants
Var. 2b = superficial fertilization-liming variant + agricultural auxiliary plants

Variant	Date of Survey	Biomass (kg ha^{-1}) Agricultural herbs		clover
	6.91	-		-
1b	8.91	white (yellow) mustard	9.5	
		mallow	2.1	
		oil radish	105.7	
		wild turnip	140.0	
		total	257.3	86.9
	10.91	oil radish	1005.9	
		wild turnip/mallow/phacelia	266.4	
		total	1272.3	143.2
	5.92	wild turnip	190.5	
		phacelia	2.0	
		total	192.5	426.6
	7.92	oil radish	35.7	
		phacelia	1.1	
		total	36.8	1200.0
	9.92	oil radish	25.0	643.3
2b	6.91	-		-
	8.91	oil radish	10.3	9.0
	10.91	wild turnip	68.4	
		oil radish	63.7	
		phacelia	6.8	
		Mallow	0.7	
		total	139.6	37.0
	5.92	Wild turnip/oil radish	56.0	
		phacelia	0.0	
		total	56.0	
	7.92	wild turnip/oil radish	51.2	193.2
	9.92	-		17.3

The clover species growing in the deep ploughing variant (Var. 1b) fix considerable amounts of the elements mentioned above during the second growing season, although their smaller biomass than that of the group of graminales. Thus, they show an ion uptake behaviour which resembles that of the group of 'herbaceous plants'.

In order to investigate whether the agricultural auxiliary plants would on principal be in a position to contribute to a more favourable nutrient storage in the biomass due to selective uptake, the maximum element contents (in mg g^{-1} dry mass) of certain species and groups of species are compared in the following. In this way it can be shown, which elements are increasingly taken up by the agricultural auxiliary plants in comparison to the dominant site-related species or groups of species. In order to accumulate an additional or "favourable" storage quantity, possibly by way of the ability to take up of certain nutrients selectively, it must, however, be assumed that the sown species will form the respective biomass. This was at least not the case on the trial plots of the superficial fertilisation-liming variant (Var. 2b). Only when it becomes apparent that the naturally occurring species are not capable of securing the formation of stored products on sufficient terms it could be advantageous to sow species 'alien to the flora`.

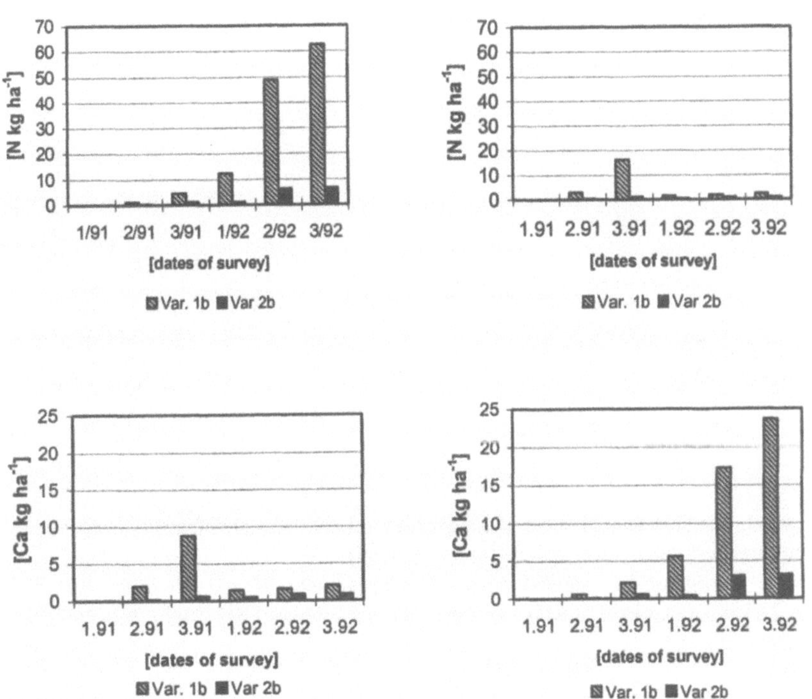

Fig. 4.3-1. Medium annual sums of nitrogen (above) and calcium (below) stored in agricultural herbs and clover species (above-ground).
(additive representation, separated into 91 and 92 growing seasons; dates: 1.91 = June 1991, 2.91 = August 1991, 3.91 = October 1991, 1.92 = May 1992, 2.92 = July 1992, 3.92 = September 1992)
Var. 1b = ploughland variant + agricultural auxiliary plants
Var. 2b = surficial fertilization variant + agricultural auxiliary plants

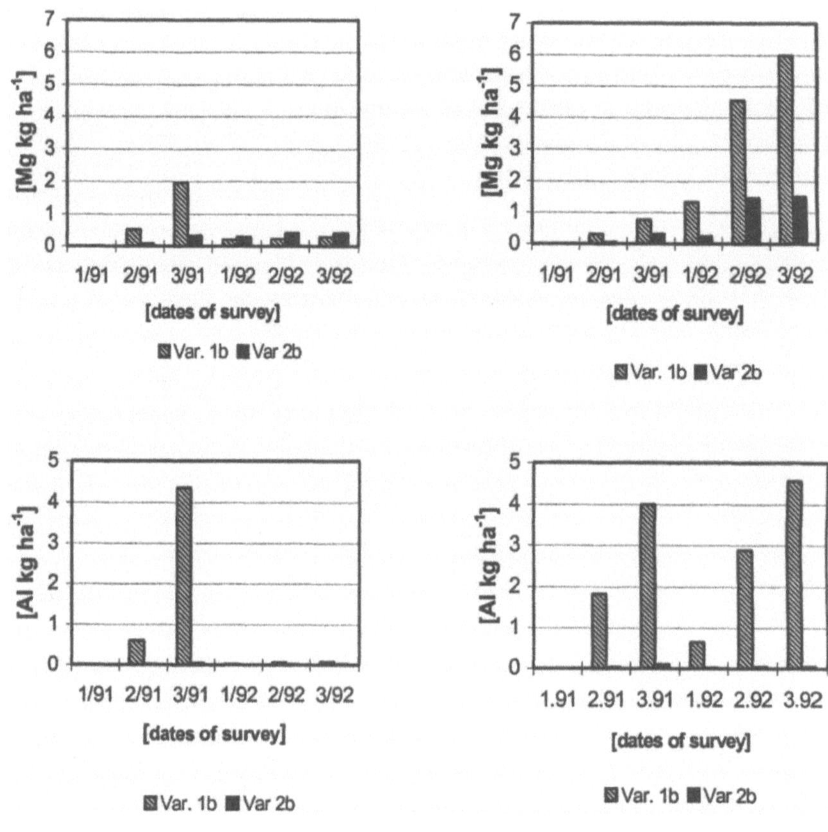

Fig. 4.3-2. Medium annual sums of magnesium (above) and aluminum (below) stored in agricultural herbs and clover species (above-ground).
(additive representation, subdivided by growing seasons 91 and 92; dates: 1.91 = June 1991, 2.91 = August 1991, 3.91 = October 1991, 1.92 = May 1992, 2.92 = July 1992, 3.92 = September 1992)

Table 4.3-6 shows the highest contents measured at all dates of the survey. A comparison is made between the "uninfluenced" variants (Linke 1994), i.e. the variants on which no agricultural auxiliary plants had been sown (deep ploughing variant without agricultural auxiliary plants = Var. 1a; superficial fertilisation-liming without agricultural auxiliary plants = Var. 2a, and control = Var. 3) and the variants with seeding of auxiliary plants (deep ploughing + agricultural auxiliary plants = Var. 1b, superficial fertilisation-liming + agricultural auxiliary plants = Var. 2b).

Table 4.3-6. Maximum element-concentrations (mg g^{-1} dry mass) of the above-ground biomass of certain species or groups of species (survey date, respectively).

Elements	Variant	Species / species groups					*Agrostis tenuis*
		Agricult. Herbs	Clover		Total biomass		
N	1b	33.0 (1.91)	30.1 (2.92)	1a	20.6 (2.91)		22.5 (3.91)
	2b	25.0 (3.91)	32.0 (1.92)	2a	21.0 (1.91)		22.5 (1.91)
				3	22.3 (3.91)		21.0 (2.91)
Ca	1b	22.0 (2.92)	13.0 (1.92)	1a	7.4 (3.91)		2.9 (1.91)
	2b	15.1 (3.91)	14.0 (2.92)	2a	5.1 (3.91)		2.3 (1.91)
				3	22.3 (3.91)		2.0 (2.92)
Mg	1b	5.2 (2.91)	3.5 (2.91)	1a	2.6 (3.91)		2.0 (3.91)
	2b	7.2 (2..91)	7.0 (3.91)	2a	3.2 (3.91)		2.0 (1.92)
				3	1.6 (2.91)		1.5 (2.91)
Al	1b	6.5 (3.91)	21.5 (2.91)	1a	8.0 (2.91)		7.0 (3.91)
	2b	1.8 (2.91)	2.5 (2.91)	2a	1.0 (2.91)		2.5 (2.91)
				3	2.6 (3.91)		3.0 (2.91)

Table 4.3-7. Maximum above-ground element-concentration (mg g^{-1} dry mass) – continuation.

Elements	Variant	Species / species groups			
		Avenella flexuosa	*Galium harciynicum*	*Carex pilulifera*	*Vaccinium myrtillus*
N	1a	22.5 (1.91)	16.5 (1.92)	30.0 (2.91)	16.0(1.92)
	2a	19.0 (1.91)	24.5 (1.91)	19.0 (1.92)	22.0(1.92)
	3	17.5 (3.91)	26.0 (3.91)	24.0 (3.91)	19.5(1.91)
Ca	1a	5.5 (3.91)3	9.0 (2.92)	4.8 (2.91)	2.9 (1.92)
	2a	2.0 (3.91)	8.4 (2.91)	3.0 (3.91)	9.3 (3.91)
	3	4.3 (3.91)	6.5 (2.91)	3.7 (3.91)	10.8(2.92)
Mg	1a	2.8 (3.92)	3.1 (2.91)	3.0 (2.91)	- -
	2a	2.2 (2.92)	4.3 (3.91)	2.1 (3.91)	2.2 (3.91)
	3	1.1 (3.91)	1.5 (3.92)	1.6 (3.91)	2.2 (2.92)
Al	1a	23.5 (3.92)	21.0 (2.91)	18.0 (2.91)	2.0 (1.92)
	2a	2.0 (2.92)	2.5 (2.91)	2.5 (2.91)	1.0 (3.91)
	3	2.1 (3.92)	4.5 (3.91)	2.2 (3.91)	0.8 (3.91)

The comparison of maximum element concentrations (s. Table 4.3-6 and 4.3.-7) reveals, that compared to those species being dominant by nature, the 'agricultural herbs' and clover species show the highest N, Ca and Mg contents – at least temporarily.

The differences to the other species or the entire above-ground biomass, respectively, are conspicuously high only in regard to calcium and magnesium. Here the 'agricultural herbs' and clover species partly appear to have more than twice as high maximum concentrations. The lowest maximum values are measured in the control variant. This allows to conclude that the sown agricultural auxiliary plants in the deep ploughing and superficial fertilisation-liming variant selectively take up a higher amount of the applied nutrients than the other dominant species and groups of species. At least the liming could be more effectively utilised by a sufficient biomass production of this species.

4.3.5 Conclusions

In summary, the investigation shows that the short-lived agricultural auxiliary plants are not able to form a nutrient-rich leaf litter on a sustainable basis. Hence, it is impossible to fix larger ion quantities in the element cycle by these auxiliary plants than by the site-related species. Thus, no additional nutrient supply is available for the crop tree species. Certain species of earthworms by which the litter is properly incorporated into the soil depend on such a sustainable, nutrient-rich litter as their nutritional basis (Wolters and Schauermann 1989).

In view of the Ca and Mg deficiency of the graminaceous plants, especially of *Agrostis tenuis* and *Avenella flexuosa*, the liming/fertilisation-liming could be more efficiently utilised by sowing species which have species-specifically higher Ca and Mg contents. This, in turn, presupposes the sustainable formation of a respective biomass. Although the agricultural auxiliary plants appear to have higher Ca and Mg contents than the dominant, site-related species of the overall above-ground biomass, they do not fulfill the preconditions of a sustainable biomass formation. As has become obvious from the investigations on the frequency and relative dominance of the species and groups of species, the agricultural auxiliary plants tend to be more disturbing for the establishment of site-related herbaceous species so that mainly graminaceous plants are found to dominate. The liming/fertilisation-liming is utilised by the latter relatively ineffectively due to their species-specific Ca and Mg deficiency, and in addition to this they are known as 'oppressor' of the planted tree species by 'encroaching upon' the crop plants in such a way that they lean on them from above or laterally and seem to 'crush' them virtually. Hence, a rapid establishment of the plots by sowing appropriate, site-related forest herbs which are capable of fulfilling the named preconditions should be taken into consideration. Experimentation with forest herbs and those thriving well on clear-cut areas is still lacking, but the following species would be conceivable: *Taraxacum officinale, Senecio sylvaticus, Moehringia trinervia, Stellaria media, Cardamine flexuosa* and others. At the same time the problem of floral distortion is avoided in this way, and colonisation by earthworm species which render a respective contribution to bioturbation inside the soil is facilitated on a sustainable basis by a nutrient-rich litter.

References

Gerlach A, Krause A, Meisel K, Speidel B, Trautmann W (1970) Vegetationsuntersuchungen im Solling. (Beitr. I.B.P. – Sollingprojekt). Schriftenreihe Vegkde:75-98

Hildebrand E (1990) Der Einfluß von Forstdüngungen auf die Lösungsfracht des Makroporenwassers. Allg Forstz 45:604-607

Linke J (1994) Untersuchungen zu Vegetationsökologie und Stoffhaushalt der Sekundärsukzession auf einer Meliorationsfläche im Solling. Bot Diss Bot 224:1-249

Londo G (1975) Dezimalskala für die vegetationskundliche Aufnahme von Dauerquadraten. Sukzessionsforschung. Ber Internat Sympos Internat Vereinig Vegkde, Rinteln 1973:613-622

Pfalz W, Irrgang S, Pohl R (1991) Das Ökologische Meßfeld der Abteilung Forstwirtschaft der TU Dresden. Teil XIII: Sukzession der Feldschicht nach Kahlhieb eines 100jährigen Fichtenaltholzes – oberirdische Phytomasse. Wiss Z Techn Univ Dresden 40:297-307

Reinecke H (1990) Müssen Kahlflächen verwildern? Aufforstung von Windwurfflächen unter Nutzpflanzendecken möglich. Allg Forstz 45:950-955.

Schieri R, Göttlein A, Hohmann E, Trübenbach D, Kreutzer K (1986) Einfluß von saurer Beregnung und Kalkung auf Humusstoffe sowie Aluminum- und Schwermetalldynamik in wäßrigen Bodenextrakten. Forstw Cbl 105:309-313

Ulrich B (1989) Forschungszentrum Waldökosysteme der Universität Göttingen – Stabilitätsbedingungen von Waldökosystemen. Forschungsantrag an das BMFT Bonn. Ber Forschungszentrum Waldökosysteme, Göttingen, Reihe B, Bd 14

Wenzel B (1989) Kalkungs- und Meliorationsexperimente im Solling: Initialeffekte auf Boden, Sickerwasser und Vegetation. Ber Forschungszentrum Waldökosysteme, Göttingen, Reihe A, Bd 51

Wolters V, Schauermann J (1989) Die Wirkung von Meliorationsmaßnahmen auf die ökologische Funktion von Lumbriciden. Ber Forschungszentrum Waldökosysteme, Göttingen, Reihe A, Bd 49

Subject index

Xp = page X and following page;
Xpp = page X and following pages

A

Abies procera 185pp
Acer platanoides 185pp
Acer pseudoplatanus 185pp
Acid rain 93
Acidification 158pp, 162, 177
Afforestation 182pp, 199pp
Agricultural auxiliary plants 199pp
Agrostis tenuis 199pp
Alnus glutinosa 169, 185pp, 201
Aluminum 76, 114, 132, 157, 176, 201, 209
Amelioration 167pp, 184
Avenella flexuosa 206, 214

B

Betula pendula 5, 22, 42, 185
Biomass
  herbaceous vegetation 121, 161, 206pp
  litter 123p
  needles 104pp
  roots 121, 161
  tree regeneration 121
Bioturbation 214
Brassica rapa spp. *oleifera* 199pp

C

Cadmium 109
Calcium 76, 100p, 157, 172pp, 192, 209pp
Calluna vulgaris 5, 23
Canopy

disintegration 143pp
  drip *see* throughfall
  interception 9, 50p
  removal 113
  throughfall 9, 23, 51, 76, 127, 155
Carbon 172p
Cation 109, 132, 144, 157p
Chronosequence 1pp
Critical loads 144
Crown vitality 93

D

De-acidification 76
Decomposer community 125
Defoliation 143
Denitrification 127
Deposition 113pp, 136
Disturbance 110, 136, 143pp
Drought 77, 80, 90pp, 106

E

Earthworm 214
Element budget 134p, 157pp
Element concentration
  in canopy throughfall 128
  in herbaceous vegetation 122, 208pp
  in leaf litter 124
  in mineral soil 147
  in leaves 98pp, 191pp
  in precipitation 127p

Mineralization 126p
Model calculation of element storage
    104

N

Needle loss 94
Needle yellowing 94
Nitrate 76, 128pp, 152pp, 177
Nitrification 127
Nitrogen 76, 97, 100, 126pp, 144,
    152pp, 159, 173pp, 191, 209pp
Nutrient cycling 93pp, 109pp,
    134pp, 156pp

O

Organic layer 21, 28, 32, 49, 55, 60,
    135, 161, 171

P

pH value 129, 152, 174
Phacelia tanacetifolia 199pp
Phosphorous 76, 99p, 191
Picea abies 69pp, 74, 80pp, 93pp,
    111, 169, 184, 200p
Pinus sylvestris 5pp
Pioneer tree species 4, 7, 60, 186
Plant growth
    height 80pp, 192
    diameter 83, 87
Plant mortality 188, 205
Ploughing 201
Potassium 97, 123, 127, 157, 191
Precipitation 11p, 76p, 127p, 155
Pseudotsuga menziesii 185

Q

Quercus petraea 5pp

R

Raphanus sativus 199pp
Rewetting 77
Roof project 69pp, 80pp, 93pp

Root water extraction 9
Roots 9, 121, 161
Rubus 120, 150, 201

S

Salix caprea 120, 189
Sambucus racemosa 150, 163
Saprophages 125
Seepage water output 130pp, 153pp,
    176pp
Sinapis alba 199pp
Soil
    evaporation 8p, 58
    fauna 125
    hydrology 1pp, 118, 155
    matric potential 7, 12, 17pp, 34,
    43pp, 116, 119
    physical properties 14
    solution 78, 115, 129, 145, 152,
    173pp
    temperature 115, 117
    water content 1pp, 22pp, 137
    water potential see soil matric
    potential
Solling 71, 73, 93, 113, 114, 169,
    183, 200
Sorbus aucuparia 185
Species diversity 111p
Stability 111
Stand structure 84, 110
Succession 1pp, 60pp, 110pp, 199,
    208
    canopy interception 50
    changes in ecosystem water
        turnover 50
    changes in soil hydrology 14
    dynamic of colonisation 202
    ecosystem water balance 58
    soil matric potential 17pp, 34
    soil water content 24
    stand evaporation 51
    throughfall 50
    water ressources 13